BROADBAND MATCHING

THEORY AND IMPLEMENTATIONS

Problems & Solutions

To Wilfred:

With my best personal wishes.

11/4/93

BROADBAND MATCHING

THEORY AND IMPLEMENTATIONS

Problems & Solutions.

World Scientific
Singapore • New Jersey • London • Hong Kong

Published by

World Scientific Publishing Co. Pte. Ltd.
P O Box 128, Farrer Road, Singapore 9128
USA office: Suite 1B, 1060 Main Street, River Edge, NJ 07661
UK office: 73 Lynton Mead, Totteridge, London N20 8DH

BROADBAND MATCHING: THEORY AND IMPLEMENTATIONS
PROBLEMS AND SOLUTIONS

ISBN 981-02-1453-7

Printed in Singapore.

v

PREFACE

The solutions to the problems in the text **Broadband Matching: Theory and Implementations** are presented in this book. Attempt has been made to conform the notation of the text, and the approach of the text is used in solving the problems. This book is intended as an aid for the instructors using the textbook, not for the use of students. It contains solutions to all the problems except those involving proofs of theorems and identities and the verification of solutions. The steps and manipulations used in the solutions are provided in details.

The solutions to the problems were worked out by graduate students, teaching assistants, and visiting scholars. Most of the solutions to the first five chapters were contributed by my doctoral students Dr. Eishi Yasui and Mr. Jian Gong. More specifically, many of the solutions to the first two chapters were provided by Dr. Yasui, and most of Chapters 3, 4 and 5 by Mr. Gong. The material was initially typed by Mr. Chi-Kuang Chao. Professor Yi-Sheng Zhu of Dalian Maritime University proofread the entire five chapters.

I wish to express my appreciation to all the persons mentioned above whose valuable contributions made the project possible.

Wai-Kai Chen   May 15, 1993
Department of Electrical Engineering
  and Computer Science
University of Illinois at Chicago
Chicago, Illinois 60680

CONTENTS

CHAPTER 1

FOUNDATIONS OF NETWORK THEORY

Problem 1.11 *Test the following function to see if it is positive-real:*

$$f(s) = \frac{2s^4 + 7s^3 + 11s^2 + 12s + 4}{s^4 + 5s^3 + 9s^2 + 11s + 6}$$

SOLUTION:

(i) $f(s)$ is real.

(ii)

$$p(s) + q(s) = 3s^4 + 12s^3 + 20s^2 + 23s + 10$$

$$\psi(s) = \frac{3s^4 + 20s^2 + 10}{12s^3 + 23s} = \frac{1}{4}s + \cfrac{1}{\cfrac{48}{57}s + \cfrac{1}{\cfrac{3249}{3324}s + \cfrac{1}{\cfrac{831}{570}s}}}$$

    (a) All α's are real and positive.

    (b) $\psi(s)$ does not terminate prematurely.

(iii)

$$A(\omega^2) = 2\left(\omega^8 + 3\omega^6 - 11\omega^4 + 15\omega^2 + 12\right)$$

$$A(x) = 2\left(x^4 + 3x^3 - 11x^2 + 15x + 12\right)$$

By Sturm's theorem, $A(x)$ has real positive roots:

$$P_0(x) = x^4 + 3x^3 - 11x^2 + 15x + 12$$

$$P_1(x) = 4x^3 + 9x^2 - 22x + 15$$

$$P_2(x) = \frac{115}{16}x^2 - \frac{123}{8}x - \frac{147}{16}$$

$$P_3(x) = -18.669x - 37.442$$

$$P_4(x) = 11.113$$

Thus $f(s)$ is P.R.

Problem 1.12 *A nonlinear time-invariant resistor is characterized by the relation*

$$v(t) = i^2(t)$$

Show that this resistor is active.

SOLUTION: Let $i = -e^t$. Then

$$\int_{-\infty}^{t} v(x)i(x)dx = \int_{-\infty}^{t} i^3(x)dx = \int_{-\infty}^{t} -e^{3x}dx = -\frac{e^{3t}}{3} < 0$$

Thus, the device is active.

Problem 1.13 *The matrix*

$$\mathbf{Y}(s) = \frac{1}{q(s)} \begin{bmatrix} 4s^2 + 12s + 1 & -1 \\ -1 & 12s^2 + 8s + 1 \end{bmatrix}$$

where

$$q(s) = 12s^3 + 44s^2 + 28s + 5$$

is known as the admittance matrix of a passive two-port network. Show that it is a positive-real matrix.

SOLUTION: The matrix $\mathbf{Y}(s)$ is positive real, because it can be realized by a passive T-network with series impedances

$$z_1 = 2 + 3s, \quad z_2 = 3 + s$$

and shunt admittance

$$y(s) = 4s$$

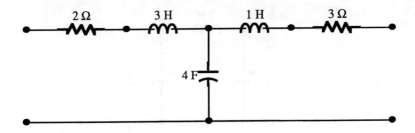

Problem 1.15 *Test the following matrices to see if they are positive-real:*

(i)

$$A(s) = \frac{1}{46s + 1} \begin{bmatrix} 4s + 2 & 4s \\ 4s - 20 & 4s + 1 \end{bmatrix}$$

(ii)

$$A(s) = \frac{1}{2s(s^2 + 1)} \begin{bmatrix} 2s^2 + 1 & 1 \\ 1 & 2s^2 + 1 \end{bmatrix}$$

(iii)

$$A(s) = \frac{1}{q(s)} \begin{bmatrix} 2s^3 + 2s^2 + 2s + 1 & -1 \\ -1 & 8s^3 + 2s^2 + 8s + 1 \end{bmatrix}$$

where

$$q(s) = 8s^3 + 10s^2 + 10s + 5$$

SOLUTION:

(i) The matrix $A(s)$ is not positive real.

(ii) The matrix $A(s)$ is positive real. It can be realized by a passive LC Π-network with two shunt admittances equal to s and one series impedance equal to $2s$.

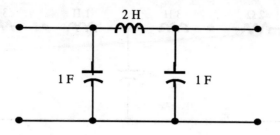

(iii) The matrix $A(s)$ is positive real, because it can be realized by an RLC two-port ladder with two shunt admittances equal to s and series impedances 4, $2s$ and 1.

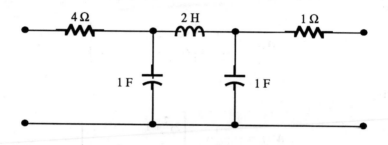

Problem 1.17 *Using Definition 1.3, show that the matrix*

$$A = \begin{bmatrix} 1 & 1 & 1 \\ 1 & 1 & 1 \\ 1 & 1 & 0 \end{bmatrix}$$

whose leading principal minors are all nonnegative, is neither positive-definite nor nonnegative-definite.

SOLUTION: Let

$$X' = \begin{bmatrix} 1 & 1 & -2 \end{bmatrix}$$

Then

$$X'AX = \begin{bmatrix} 1 & 1 & -2 \end{bmatrix} \begin{bmatrix} 1 & 1 & 1 \\ 1 & 1 & 1 \\ 1 & 1 & 0 \end{bmatrix} \begin{bmatrix} 1 \\ 1 \\ -2 \end{bmatrix} = -4 < 0$$

CHAPTER 2

THE SCATTERING MATRIX

Problem 2.3 *Consider the ideal transformer N of Fig. 2.7. Determine its scattering matrix normalized to the reference impedance matrix as given in (2.118a).*

SOLUTION:

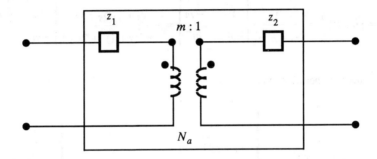

The reference impedance matrix $\mathbf{z}(s)$ is found to be

$$\mathbf{z}(s) = \begin{bmatrix} \dfrac{1}{s+1} & 0 \\ 0 & \dfrac{2s+2}{2s+1} \end{bmatrix}$$

The augmented admittance matrix is obtained as

$$\mathbf{Y}_a = \begin{bmatrix} y_{11a} & y_{12a} \\ y_{21a} & y_{22a} \end{bmatrix}$$

where

$$y_{11a} = \frac{1}{z_1 + m^2 z_2}$$

$$y_{22a} = \frac{1}{\dfrac{z_1}{m^2} + z_2} = \frac{m^2}{z_1 + m^2 z_2}$$

$$y_{21a} = \frac{I_2}{v_1}\bigg|_{v_2=0} = -\frac{mI_1}{v_1}\bigg|_{v_2=0} = -my_{11a} = \frac{-m}{z_1 + m^2 z_2} = y_{12a}$$

giving

$$\mathbf{Y}_a = [\mathbf{Z} + \mathbf{z}]^{-1} = \frac{1}{z_1 + m^2 z_2}\begin{bmatrix} 1 & -m \\ -m & m^2 \end{bmatrix} = \frac{(s+1)(2s+1)}{2m^2(s+1)^2 + 2s + 1}\begin{bmatrix} 1 & -m \\ -m & m^2 \end{bmatrix}$$

Compute the para-hermitian part of $\mathbf{z}(s)$:

$$\mathbf{r}(s) = \frac{1}{2}\big[\mathbf{z}(s) + \mathbf{z}_*(s)\big] = \begin{bmatrix} \dfrac{1}{(1+s)(1-s)} & 0 \\ 0 & \dfrac{2(1-2s^2)}{(1+2s)(1-2s)} \end{bmatrix} = \mathbf{h}(s)\mathbf{h}_*(s)$$

where

$$\mathbf{h}(s) = \begin{bmatrix} \dfrac{1}{1+s} & 0 \\ 0 & \dfrac{\sqrt{2}(1-\sqrt{2}s)}{1+2s} \end{bmatrix}$$

$$\mathbf{h}_*(s) = \begin{bmatrix} \dfrac{1}{1-s} & 0 \\ 0 & \dfrac{\sqrt{2}(1+\sqrt{2}s)}{1-2s} \end{bmatrix}$$

$$\mathbf{h}_*^{-1}(s) = \begin{bmatrix} 1-s & 0 \\ 0 & \dfrac{1-2s}{\sqrt{2}(1+\sqrt{2}s)} \end{bmatrix}$$

$$\mathbf{S}' = \mathbf{U} - 2[\mathbf{Z}+\mathbf{z}]^{-1}\mathbf{r} = \begin{bmatrix} 1-\dfrac{2(2s+1)}{q(1-s)} & \dfrac{4m(1+\sqrt{2}s)(1-\sqrt{2}s)(s+1)}{q(1-2s)} \\ \dfrac{2m(2s+1)}{q(1-s)} & 1-\dfrac{4(s+1)m^2(1+\sqrt{2}s)(1-\sqrt{2}s)}{q(1-2s)} \end{bmatrix}$$

where

$$q(s) = 2m^2(s+1)^2 + 2s + 1$$

obtaining

$$\mathbf{S} = \mathbf{h}\mathbf{S}'\mathbf{h}_*^{-1} = \begin{bmatrix} \dfrac{h_1}{h_{1*}}S'_{11} & \dfrac{h_1}{h_{2*}}S'_{12} \\ \dfrac{h_2}{h_{1*}}S'_{21} & \dfrac{h_2}{h_{2*}}S'_{22} \end{bmatrix}$$

where

$$S_{11} = \frac{\frac{1}{1+s}}{\frac{1}{1-s}} \times \left(1 - \frac{2(2s+1)}{q(1-s)}\right) = \frac{2m^2(1-s^2)-(2s+1)}{q}$$

$$S_{12} = \frac{h_1}{h_{2*}}S'_{12} = \frac{2\sqrt{2}m(1-\sqrt{2}s)}{q}$$

$$S_{21} = \frac{h_2}{h_{1*}}S'_{21} = \frac{2\sqrt{2}m(1-\sqrt{2}s)}{q}$$

$$S_{22} = \frac{h_2}{h_{2*}}S'_{22} = \frac{\sqrt{2}(1-\sqrt{2}s)(1-2s)}{(1+2s)\sqrt{2}(1+\sqrt{2}s)}\left[1 - \frac{4(s+1)m^2(1+\sqrt{2}s)(1-\sqrt{2}s)}{q(1-2s)}\right]$$

$$= \frac{1 - \sqrt{2}s}{1 + \sqrt{2}s} \times \frac{2m^2(s^2 - 1) + 1 - 2s}{q}$$

Finally, the normalized scattering matrix is found to be

$$S = \frac{1}{q}\begin{bmatrix} 2m^2(1 - s^2) - (2s + 1) & 2\sqrt{2}m\left(1 - \sqrt{2}s\right) \\ 2\sqrt{2}m\left(1 - \sqrt{2}s\right) & \dfrac{(1 - \sqrt{2}s)\left[2m^2(s^2 - 1) + (1 - 2s)\right]}{1 + \sqrt{2}s} \end{bmatrix}$$

Alternatively, we have

$$S_{11} = \frac{h_1(Z_{11} - z_{1*})}{h_{1*}(Z_{11} + z_1)} = \frac{2m^2(1 - s^2) - (2s + 1)}{q(s)}$$

$$S_{22} = \frac{h_2(Z_{22} - z_{2*})}{h_{2*}(Z_{22} + z_2)} = \frac{1 - \sqrt{2}s}{1 + \sqrt{2}s} \times \frac{1 - 2s - 2m^2(1 - s^2)}{q(s)}$$

$$S_{21} = -2h_2 h_1 \frac{I_2}{V_{g1}}$$

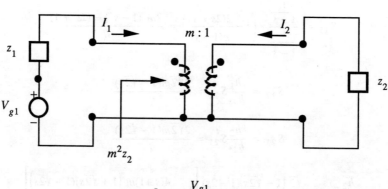

$$I_1 = \frac{V_{g1}}{z_1 + m^2 z_2}$$

$$I_2 = -mI_1 = -\frac{mV_{g1}}{z_1 + m^2 z_2}$$

$$\frac{I_2}{V_{g1}} = \frac{-m}{z_1 + m^2 z_2}$$

$$S_{21} = -2\frac{\sqrt{2}\left(1-\sqrt{2}s\right)}{1+2s} \times \frac{1}{1+s} \times \frac{-m}{\frac{1}{s+1} + m^2\frac{2s+2}{2s+1}}$$

$$= \frac{2\sqrt{2}\left(1-\sqrt{2}s\right)m}{(1+2s) + m^2(2s+2)(1+s)} = S_{12}$$

Problem 2.8 *Confirm that the scattering matrix of the lossless two-port network of Fig. 1.8, normalizing to the resistances $z_1 = 4\,\Omega$ and $z_2 = 1\,\Omega$, is given by*

$$S(s) = \frac{1}{q(s)}\begin{bmatrix} -8s^3 - 6s^2 - 6s - 3 & 4 \\ 4 & -8s^3 + 6s^2 - 6s + 3 \end{bmatrix}$$

where

$$q(s) = 8s^3 + 10s^2 + 10s + 5$$

SOLUTION: We have

$$\mathbf{z}(s) = \begin{bmatrix} 4 & 0 \\ 0 & 1 \end{bmatrix}, \qquad \mathbf{z}_*(s) = \begin{bmatrix} 4 & 0 \\ 0 & 1 \end{bmatrix}$$

$$\mathbf{Z}(s) = \frac{1}{2s(s^2+1)}\begin{bmatrix} 2s^2 + 1 & 1 \\ 1 & 2s^2 + 1 \end{bmatrix}$$

$$\mathbf{Z}(s) + \mathbf{z}(s) = \frac{1}{2s(s^2 + 1)} \begin{bmatrix} 8s^3 + 2s^2 + 8s + 1 & 1 \\ 1 & 2s^3 + 2s^2 + 2s + 1 \end{bmatrix}$$

$$\det[\mathbf{Z}(s) + \mathbf{z}(s)] = \frac{2s\left(8s^5 + 10s^4 + 18s^3 + 15s^2 + 10s + 5\right)}{\left[2s(s^2 + 1)\right]^2}$$

$$[\mathbf{Z}(s) + \mathbf{z}(s)]^{-1} = \frac{1}{q(s)} \begin{bmatrix} 2s^3 + 2s^2 + 2s + 1 & -1 \\ -1 & 8s^3 + 2s^2 + 8s + 1 \end{bmatrix}$$

where

$$q(s) = 8s^3 + 10s^2 + 10s + 5$$

As a check, we compute the admittance matrix $\mathbf{Y}_{a1} = [\mathbf{Z} + \mathbf{z}]^{-1}$ of the augmented network:

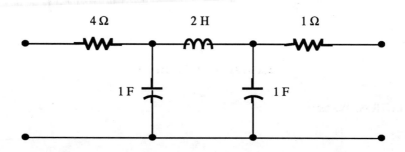

$$y_{11} = \cfrac{1}{4 + \cfrac{1}{s + \cfrac{1}{2s + \cfrac{1}{s+1}}}} = \frac{2s^3 + 2s^2 + 2s + 1}{8s^3 + 10s^2 + 10s + 5}$$

$$y_{22} = \cfrac{1}{1 + \cfrac{1}{s + \cfrac{1}{2s + \cfrac{1}{s + \cfrac{1}{4}}}}} = \frac{8s^3 + 2s^2 + 8s + 1}{8s^3 + 10s^2 + 10s + 5}$$

$$y_{21} = \frac{I_2}{V_1}\Big|_{V_2=0}$$

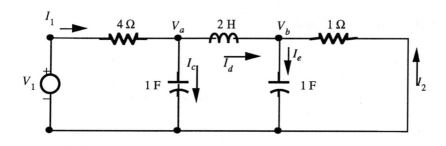

Let $I_2 = 1$. Then

$$V_b = -1$$

$$I_e = sV_b = -s$$

$$I_d = I_e - I_2 = -s - 1$$

$$V_a = 2s(-s-1) - 1 = -2s^2 - 2s - 1$$

$$I_c = -2s^3 - 2s^2 - s$$

$$I_1 = I_c + I_d = -2s^3 - 2s^2 - 2s - 1$$

$$V_1 = 4I_1 + V_a = -8s^3 - 10s^2 - 10s - 5$$

yielding

$$y_{21} = \frac{1}{V_1} = \frac{-1}{8s^3 + 10s^2 + 10s + 5} = y_{12}$$

We next compute

$$\mathbf{Z}(s) - \mathbf{z}_*(s) = \frac{1}{2s(2s^2 + 1)} \begin{bmatrix} -8s^3 + 2s^2 - 8s + 1 & 1 \\ 1 & -2s^3 + 2s^2 - 2s + 1 \end{bmatrix}$$

$$\mathbf{S}'(s) = \left[\mathbf{Z}(s) + \mathbf{z}(s)\right]^{-1}\left[\mathbf{Z}(s) - \mathbf{z}_*(s)\right]$$

$$= \frac{1}{q(s)} \begin{bmatrix} -8s^3 - 6s^2 - 6s - 3 & 2 \\ 8 & -8s^3 + 6s^2 - 6s + 3 \end{bmatrix}$$

$$\mathbf{r}(s) - \frac{1}{2}\left[\mathbf{Z}(s) + \mathbf{z}_*(s)\right] = \begin{bmatrix} 4 & 0 \\ 0 & 1 \end{bmatrix} = \mathbf{h}(s)\mathbf{h}_*(s)$$

$$\mathbf{h}(s) = \begin{bmatrix} 2 & 0 \\ 0 & 1 \end{bmatrix}, \qquad \mathbf{h}_*(s) = \begin{bmatrix} 2 & 0 \\ 0 & 1 \end{bmatrix}$$

$$\mathbf{h}_*^{-1} = \begin{bmatrix} \frac{1}{2} & 0 \\ 0 & 1 \end{bmatrix}$$

$$\mathbf{S}(s) = \mathbf{h}(s)\mathbf{S}'(s)\mathbf{h}_*^{-1}(s) = \frac{1}{q(s)} \begin{bmatrix} -(8s^3 + 6s^2 + 6s + 3) & 4 \\ 4 & -(8s^3 - 6s^2 + 6s - 3) \end{bmatrix}$$

We now compute $\mathbf{S}(s)$ using

$$S_{11}(s) = \frac{h_1(s)}{h_{1*}(s)} \times \frac{Z_{11}(s) - z_{1*}(s)}{Z_{11}(s) + z_1(s)}$$

$$S_{22}(s) = \frac{h_2(s)}{h_{2*}(s)} \times \frac{Z_{22}(s) - z_{2*}(s)}{Z_{22}(s) + z_2(s)}$$

$$S_{12}(s) = -2h_1 h_2 \frac{I_1}{V_{g2}}$$

$$S_{21}(s) = -2h_2 h_1 \frac{I_2}{V_{g1}}$$

Problem 2.13 *Using (2.104c), compute the scattering matrix of an ideal gyrator with gyration resistance of 1 Ω, normalized to the load impedances as given in (2.118a).*

SOLUTION:

$$\mathbf{Z}(s) = \begin{bmatrix} 0 & -1 \\ 1 & 0 \end{bmatrix}, \qquad \mathbf{Y}(s) = \begin{bmatrix} 0 & 1 \\ -1 & 0 \end{bmatrix}$$

$$\mathbf{z}(s) = \begin{bmatrix} \dfrac{1}{s+1} & 0 \\ 0 & \dfrac{2s+2}{2s+1} \end{bmatrix}$$

$$\mathbf{y}(s) = \begin{bmatrix} s+1 & 0 \\ 0 & \dfrac{2s+1}{2s+2} \end{bmatrix}$$

$$\mathbf{g}(s) = \frac{1}{2}\left[\mathbf{y}(s) + \mathbf{y}_*(s)\right] = \begin{bmatrix} 1 & 0 \\ 0 & \dfrac{1-2s^2}{1-s^2} \end{bmatrix}$$

$$= \begin{bmatrix} 1 & 0 \\ 0 & \dfrac{1 - \sqrt{2}s}{1 + s} \end{bmatrix} \begin{bmatrix} 1 & 0 \\ 0 & \dfrac{1 + \sqrt{2}s}{1 - s} \end{bmatrix}$$

$$\mathbf{k}(s) = \begin{bmatrix} 1 & 0 \\ 0 & \dfrac{1 - \sqrt{2}s}{1 + s} \end{bmatrix}$$

$$\mathbf{Y}(s) + \mathbf{y}(s) = \begin{bmatrix} s + 1 & 1 \\ -1 & \dfrac{2s + 1}{2s + 2} \end{bmatrix}$$

$$\det\left[\mathbf{Y}(s) + \mathbf{y}(s)\right] = \frac{1}{2}(2s + 1) + 1 = s + 1.5$$

$$\left[\mathbf{Y}(s) + \mathbf{y}(s)\right]^{-1} = \frac{1}{s + 1.5} \begin{bmatrix} \dfrac{s + 0.5}{s + 1} & -1 \\ 1 & s + 1 \end{bmatrix}$$

$$\mathbf{S}(s) = -\mathbf{k}(s)\mathbf{k}_*^{-1}(s) + 2\mathbf{k}(s)\left[\mathbf{Y}(s) + \mathbf{y}(s)\right]^{-1}\mathbf{k}(s)$$

$$= \begin{bmatrix} -1 + \dfrac{2s + 1}{(s + 1.5)(s + 1)} & \dfrac{2(\sqrt{2}s - 1)}{(s + 1)(s + 1.5)} \\ \dfrac{2(1 - \sqrt{2}s)}{(s + 1)(s + 1.5)} & \dfrac{(s - 1)(1 - \sqrt{2}s)}{(s + 1)(1 + \sqrt{2}s)} + \dfrac{2(1 - 2s^2)}{(s + 1)(s + 1.5)} \end{bmatrix}$$

$$= \frac{1}{s^2 + 2.5s + 1.5} \begin{bmatrix} -s^2 - 0.5s - 0.5 & 2\sqrt{2}s - 2 \\ 2 - 2\sqrt{2}s & \dfrac{(s+1.5)(1-\sqrt{2}s)(s-1) + 2(1+\sqrt{2}s)(1-2s^2)}{1+\sqrt{2}s} \end{bmatrix}$$

$$= \frac{1}{s^2 + 2.5s + 1.5} \begin{bmatrix} -(s^2 + 0.5s + 0.5) & 2\sqrt{2}s - 2 \\ 2 - 2\sqrt{2}s & \dfrac{1-\sqrt{2}s}{1+\sqrt{2}s}\left[5s^5 + (4\sqrt{2} + 0.5)s + 0.5\right] \end{bmatrix}$$

Problem 2.24 *A lossless reciprocal three-port network is characterized by its normalized scattering matrix:*

$$\mathbf{S} = \begin{bmatrix} \alpha & \beta & \beta \\ \beta & \alpha & \beta \\ \beta & \beta & \alpha \end{bmatrix}$$

α *being real. Prove that* $\alpha \neq 0$, *and determine the minimum* α.

SOLUTION: Since $\mathbf{S}$ is para-unitary, we have

$$\alpha^2 + 2|\beta|^2 = 1$$

$$2\alpha \, \mathrm{Re}\, \beta + |\beta|^2 = 0$$

If $\alpha = 0$, then $|\beta| = 1$ and $|\beta| = 0$. This is impossible. Thus, we have

$$\alpha \neq 0, \quad \alpha = \pm \sqrt{1 - 2|\beta|^2}$$

$$\Rightarrow \quad |\beta| \leq \frac{1}{2} \text{ and } -1 \leq \alpha \leq 1$$

$$\alpha_{min} = -1$$

Two cases are considered. **Case 1.** If $\beta = jb$, then $|b| = 0$.

$$\Rightarrow \beta = 0 \text{ and } \alpha = \pm 1$$

Case 2. If $\beta = a$, being real, $(2\alpha \pm a)a = 0$

$$\Rightarrow \quad a = 0 \text{ and } 2\alpha = \pm a$$

and

$$a = 0 \quad \Rightarrow \quad \alpha = \pm 1$$

$$a = -2\alpha \quad \Rightarrow \quad \alpha^2 + 8\alpha^2 = 1$$

or

$$\alpha = \pm \frac{1}{\sqrt{9}} = \pm \frac{1}{3}$$

$$a = \pm \frac{2}{3}$$

CHAPTER 3

APPROXIMATION AND LADDER REALIZATION

Problem 3.6 *Design a low-pass filter having a maximally-flat transducer power-gain characteristic and operating between a resistive generator of internal resistance 70 Ω and a 200-Ω load. The filter must give at least 50 dB attenuation in gain at the frequency five times the radian cutoff frequency* $\omega_c = 10^5$ *rad/s and beyond, and has a maximum permissible dc gain.*

SOLUTION:

$$n \geq \frac{\log (10^5 - 1)}{2 \times \log 5} = 3.577$$

Choose $n = 4$.

$$R_2 > R_1, \quad S_{11}(0) > 0, \quad \frac{R_2}{R_1} = \frac{200}{70} = \frac{20}{7} = R$$

$$\delta^4 = \frac{R-1}{R+1} = \frac{13}{27}, \quad K_4 = 1 - \delta^8 = 0.7682$$

$$\delta = 0.833, \quad \gamma_m = m\frac{\pi}{8} = m \times 22.5°, \quad m = 1, 2$$

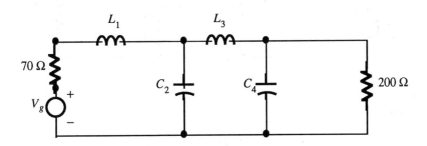

The element values are found to be

$$L_1 = \frac{2 \times 70 \sin 22.5°}{(1 - 0.833)10^5} = 3.2081 \text{ mH}$$

$$C_2 = \frac{4 \sin 22.5° \sin 67.5°}{L_1(1.6939 - 1.666 \cos 45°) \, 10^{10}} = 0.085456 \, \mu F$$

$$L_3 = \frac{4 \sin 67.5° \sin 112.5°}{C_2(1.6939 - 1.666 \cos 90°)10^{10}} = 2.3587 \, mH$$

$$C_4 = \frac{2 \sin 22.5°}{200(1 + 0.833) \, 10^5} = 0.020877 \, \mu F$$

Problem 3.7 *In Problem 3.6, suppose that the load resistance is not specified. Design a Butterworth LC ladder filter having a maximum dc gain.*

SOLUTION:

$$K_n = 1, \qquad \delta = 0, \qquad R_1 = R_2 = 70 \, \Omega$$

Choose $n = 4$. Then

$$\gamma_m = m \times 22.5°, \qquad m = 1, 2$$

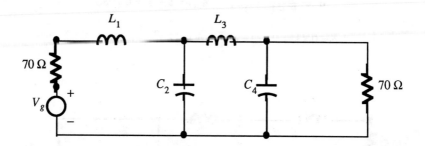

We obtain

$$L_1 = \frac{2 \times 70 \sin 22.5°}{10^5} = 0.53576 \, mH$$

$$C_2 = \frac{2 \sin 67.5°}{70 \times 10^5} = 0.264 \, \mu F$$

$$L_3 = \frac{2\times70 \sin 112.5^\circ}{10^5} = 1.2934 \text{ mH}$$

$$C_4 = \frac{2 \sin 157.5^\circ}{70\times10^5} = 0.1093 \text{ } \mu\text{F}$$

Problem 3.15 Repeat Problem 3.6 for a Chebyshev transducer power-gain characteristic having a 1-dB peak-to-peak ripple in the passband.

SOLUTION:

$$\varepsilon = \sqrt{10^{0.1} - 1} = 0.5088$$

$$n \geq \frac{\frac{1}{2} \ln \frac{4\times10^5}{0.5088}}{\cosh^{-1} 5} = 2.96$$

Thus, $n = 3$, where

$$\cosh^{-1} 5 = \ln\left(5 + \sqrt{5^2 - 1}\right) = 2.2924$$

We next compute

$$\frac{R_2}{R_1} = \frac{200}{70} = \frac{1 + (1 - K_3)^{1/2}}{1 - (1 - K_3)^{1/2}} \quad \rightarrow \quad K_3 = 1 - \left[\frac{\frac{R_2}{R_1} - 1}{\frac{R_2}{R_1} + 1}\right]^2 = 0.768$$

$$\gamma_m = m \frac{\pi}{2n} = m \times 30^\circ$$

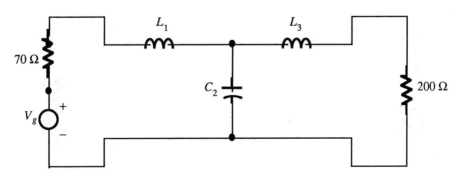

$$a = \frac{1}{3}\sinh^{-1}\frac{1}{0.5088} = 0.476$$

$$\hat{a} = \frac{1}{3}\sinh^{-1}\frac{\sqrt{1-0.768}}{\varepsilon} = 0.281$$

$$L_1 = \frac{2\times70\sin 30°}{10^5(\sinh 0.476 - \sinh 0.281)} = 3.342\ \text{mH}$$

$$C_2 = \frac{4\times\sin 30°\sin 90°}{L_1 10^{10}\left(\sinh^2 a + \sinh^2 \hat{a} + \sin^2 60° - 2\sinh a\ \sinh \hat{a}\ \cos 60°\right)} = 0.064\ \mu\text{F}$$

$$L_3 = \frac{2\times200\sin 30°}{10^5(\sinh a + \sinh \hat{a})} = 2.568\ \text{mH}$$

Problem 3.16 *In Problem 3.15, suppose that the load resistance is not specified. Design a Chebyshev LC ladder filter having a maximum attainable K_n.*

SOLUTION: Given $R_1 = 70\ \Omega$, $\omega_c = 10^5$ rad/s, $A_{dB} \geq 50$ dB, and the ripple is 1-dB. The maximum attainable $K_3 = 1$ if $R_2 = R_1 = 70\ \Omega$. We have

$$\varepsilon = 0.5088, \quad n = 3, \quad a = 0.476$$

$$K_3 = 1 \quad \rightarrow \quad \hat{a} = 0 \quad \rightarrow \quad \sinh \hat{a} = 0$$

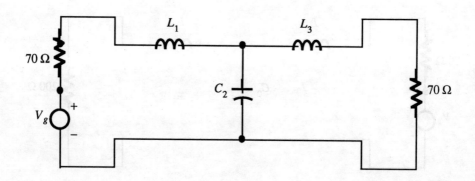

$$L_1 = \frac{2 \times 70 \sin 30°}{10^5 \sinh a} = 1.416 \text{ mH} = L_3$$

$$C_2 = \frac{4 \sin 30° \sin 90°}{L_1 10^{10}\left(\sinh^2 a + \sinh^2 60°\right)} = 0.142 \text{ μF}$$

Problem 3.17 *Design a low-pass Chebyshev filter having the following specifications:*

(i) *Peak-to-peak ripple in the passband must not exceed 1.5 dB.*
(ii) *The minimum attenuation at three times the cutoff frequency, which is 50 MHz, and beyond is 40 dB.*
(iii) *A resistive generator of internal resistance 150 Ω is the excitation, and the load resistance is 470 Ω.*

SOLUTION:

$$\omega_c = 2\pi f_c = 3.14159 \times 10^8 = 10^8 \pi \text{ rad/s}$$

$$\varepsilon = \sqrt{10^{0.15} - 1} = 0.64229$$

$$n \geq \frac{\frac{1}{2} \ln \frac{4 \times 10^4}{0.64229}}{\cosh^{-1} 3} = 3.13 \quad \rightarrow \quad n = 4$$

$$G_{min} = 1 - \left[\frac{\frac{470}{150} - 1}{\frac{470}{150} + 1}\right]^2 = 0.7336$$

$$K_4 = G_{min}\left(1 + \varepsilon^2\right) = 1.036 > 1$$

Let $K_4 = 1$.

$$\varepsilon = \sqrt{\frac{1}{0.7336} - 1} = 0.6026 \quad \rightarrow \quad 1.345 \text{ dB} \leq 1.5 \text{ dB}$$

$$\gamma_m = m\frac{\pi}{2 \times 4} = 22.5°m$$

22

$$a = \frac{1}{4} \sinh^{-1} \frac{1}{\varepsilon} = 0.32, \qquad \hat{a} = 0$$

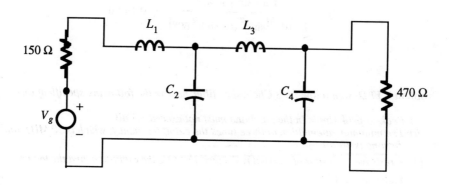

$$L_1 = \frac{2 \times 150 \sin 22.5°}{\pi 10^8 \sinh 0.32} = 1.123 \ \mu H$$

$$C_2 = \frac{4 \sin 22.5° \sin 67.5°}{L_1 \pi^2 10^{16} (\sinh^2 a + \sin^2 45°)} = 21.062 \ pF$$

$$L_3 = \frac{4 \sin 67.5° \sin 112.5°}{C_2 \pi^2 10^{16} (\sinh^2 a + \sin^2 90°)} = 1.485 \ \mu F$$

$$C_4 = \frac{2 \sin 22.5°}{470 \pi 10^8 \sinh a} = 15.92 \ pH$$

Problem 3.22 *Repeat the problem stated in Example 3.3 for a Chebyshev transducer power-gain characteristic having a 3-dB peak-to-peak ripple in the passband. Compare your result with the Butterworth case.*

SOLUTION:

$$\varepsilon = \sqrt{10^{0.1 \times 3} - 1} = 0.9976$$

$$n \geq \frac{\frac{1}{2} \ln \frac{4 \times 10^6}{0.9976}}{\cosh^{-1} 5} = 3.316 \quad \rightarrow \quad n = 4$$

$$K_4 > 1 \quad \rightarrow \quad K_4 = 1 \quad \rightarrow \quad \varepsilon = 0.3536$$

The ripple is 0.5115 dB < 3 dB.

$$\sinh a = 0.4551, \qquad \sinh \hat{a} = 0$$

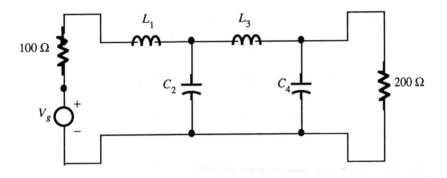

$$L_1 = 16.8179 \text{ mH}, \qquad C_2 = 1.1892 \, \mu\text{F}$$

$$L_3 = 23.7841 \text{ mH}, \qquad C_4 = 0.8409 \, \mu\text{F}$$

Comparing the results with the Butterworth case (Example 3.3), we find that

(1) There is one less component in the Chebyshev low-pass filter than that of the Butterworth one, being one less order.

(2) The Chebyshev response has the same ripple in the passband, whereas the Butterworth response decreased monotonously.

Problem 3.23 *Repeat Problem 3.17 for 1 dB ripple in the passband.*

SOLUTION:

$$n = 4, \qquad \varepsilon = 0.5088, \qquad K_4 = 0.92356$$

$$\sinh a = 0.3646, \qquad \sinh \hat{a} = 0.1303$$

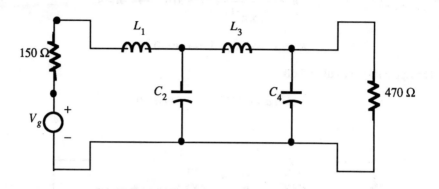

$$L_1 = 1.5594 \text{ mH}, \qquad C_2 = 15.7686 \text{ μF}$$

$$L_3 = 1.9078 \text{ mH}, \qquad C_4 = 10.4739 \text{ μF}$$

Problem 3.30 *Repeat the problem given in Example 3.6 for a passband tolerance of 1/2 dB.*

SOLUTION:

$$n = 4, \qquad \varepsilon = 0.3493, \qquad K_4 = 0.9973$$

$$\sinh a = 0.4582, \qquad \sinh \hat{a} = 0.03672$$

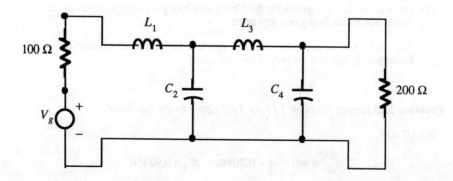

$$L_1 = 18.1582 \text{ mH}, \quad C_2 = 1.1328 \text{ } \mu\text{F}$$

$$L_3 = 24.8815 \text{ mH}, \quad C_4 = 0.7732 \text{ } \mu\text{F}$$

Problem 3.37 *Determine the Hurwitz polynomial (3.231) for n = 3, 1/k = 1.25 and ε = 0.35.*

SOLUTION:

$$k = \frac{1}{1.25} = 0.8, \quad \sin^{-1} k = 53.13°$$

$$K = F\left(k, \frac{\pi}{2}\right) = \frac{1.9927 - 1.9729}{1°}(53.13° - 53°) + 1.9729 = 1.9755$$

$$\omega_1 = \text{sn}\left(\frac{2K}{n}, k\right) = \text{sn}\left(\frac{2 \times 1.9755}{3}, 0.8\right) = \text{sn}(1.31698, 0.8)$$

$$= \sin 65.42988° = 0.90945$$

$$k_1 = k^3\left[\frac{1-\omega_1^2}{1-k^2\omega_1^2}\right]^2 = (0.8)^3\left[\frac{1-0.90945^2}{1-(0.8)^2 \times 0.90945^2}\right]^2 = 0.069092$$

$$k_1' = \sqrt{1-k_1^2} = 0.99761, \quad K_1 = F\left(k_1, \frac{\pi}{2}\right) = 1.5727$$

$$a = -j\frac{K}{nK_1}\text{sn}^{-1}\left(\frac{j}{\varepsilon}, k_1\right) = \frac{K}{nK_1}F\left[\sin^{-1}\left(1+\varepsilon^2\right)^{-1/2}, k_1'\right]$$

$$= \frac{1.9755}{3 \times 1.5727}F(\sin^{-1}0.943858, 0.99761)$$

$$= 0.41871 \times 1.7575 = 0.735895$$

$$Y_{p_0} = j\,\text{sn}(j0.735895, 0.8) = -\text{tn}(0.735895, 0.6) = -\tan 40.91° = -0.86647$$

$$Y_{p1}, Y_{p(-1)} = j\,\text{sn}\left(\pm\frac{2K}{3} + ja, k\right) = j\,\text{sn}(\pm 1.317 + j0.735895, 0.8)$$

$$= -0.17577 \pm j1.04847$$

where

$$k' = \sqrt{1-k^2} = 0.6, \qquad \text{am}\,(u, k) = 65.56°, \qquad \text{am}\,(v, k') = 40.94°$$

and (3.153a) applies, obtaining

$$r(y) = (y - y_{p}{}_{0})(y - y_{p1})(y - \bar{y}_{p1}) = y^3 + 1.2180y^2 + 1.4348y + 0.9793$$

Problem 3.39 *Repeat problem 3.37 for n = 4.*

SOLUTION:

$$k = 1.9755, \qquad k' = 0.6$$

$$\omega_1 = \text{sn}\left(\frac{K}{n}, k\right) = \text{sn}\,(0.493875, 0.8) = \sin 27.6° = 0.463296$$

$$\omega_2 = \text{sn}\left(\frac{3K}{n}, k\right) = \text{sn}\,(1.481625, 0.8) = \sin 71.85° = 0.950244$$

$$k_1 = k^4 \left[\frac{(1 - \omega_1^2)(1 - \omega_2^2)}{(1 - k^2\omega_1^2)(1 - k^2\omega_2^2)} \right]^2 = 0.017942$$

$$k_1' = \sqrt{1 - k_1^2} = 0.999839, \qquad K_1 = F\left(k_1, \frac{\pi}{2}\right) = 1.57093$$

$$a = -j\frac{K}{nK_1}\,\text{sn}^{-1}\left(\frac{j}{\varepsilon}, k_1\right) = 0.31438\,F(70.71°, 0.999839)$$

$$= 0.31438 \times 1.771726 = 0.557$$

$$y_{p1},\ y_{p(-1)} = j\,\text{sn}\left(\pm\frac{3K}{n} + ja, k\right) = j\,\text{sn}\,(\pm 1.481625 + j0.557, 0.8)$$

$$= -0.10154 \pm j1.01925$$

where

$$\text{am}\,(u, k) = 71.85°, \qquad \text{am}\,(v, k') = 31.36°$$

$$y_{p2},\ y_{p(-2)} = j\,\text{sn}\left(\pm\frac{K}{n} + ja, k\right) = j\,\text{sn}\,(\pm 0.493875 + j0.557, 0.8)$$

$$= -0.47727 \pm j0.57430$$

where

$$am\ (u, k) = 27.6°, \quad am\ (v, k') = 31.36°$$

$$r\ (y) = y^4 + 1.1576y^3 + 1.8006y^2 + 1.1147y + 0.5850$$

Problem 3.40 *Determine the characteristic function for n = 3 and 1/k = 1.1.*

SOLUTION:

$$k = 0.90909, \quad K = F\left(k, \frac{\pi}{2}\right) = 2.321924$$

$$\omega_1 = sn\left(\frac{2K}{n}, k\right) = sn\ (1.54295, 0.90909) = \sin 70° = 0.93969$$

$$k_1 = k^3\left[\frac{1 - \omega_1^2}{1 - k^2\omega_1^2}\right]^2 = 0.140787$$

$$H_0 = \sqrt{\frac{k^3}{k_1}} = 2.31009$$

$$F_3(\omega) = H_0\frac{\omega(\omega_1^2 - \omega^2)}{1 - k^2\omega_1^2\omega^2} = 2.31009\frac{\omega(0.88302 - \omega^2)}{1 - 0.72977\omega^2}$$

Problem 3.41 *Determine the characteristic function for n = 4 and 1/k = 1.1.*

SOLUTION:

$$k = 0.90909, \quad K = K(k) = 2.321924$$

$$\omega_1 = sn\left(\frac{K}{n}, k\right) = sn\ (0.580481, 0.90909) = \sin 31.83° = 0.52740$$

$$\omega_2 = sn\left(\frac{3K}{n}, k\right) = sn\ (1.741443, 0.90909) = \sin 75.5° = 0.96815$$

$$k_1 = k^4 \left[\frac{(1 - \omega_1^2)(1 - \omega_2^2)}{(1 - k^2\omega_1^2)(1 - k^2\omega_2^2)} \right]^2 = 0.683013 \left[\frac{(1 - 0.27815)(1 - 0.93731)}{(1 - 0.22988)(1 - 0.77464)} \right]^2$$

$$= 0.046433$$

$$H_e = \sqrt{\frac{k^4}{k_1}} = 3.8353$$

$$F_A(\omega) = H_e \frac{(\omega_1^2 - \omega^2)(\omega_2^2 - \omega^2)}{(1 - k^2\omega_1^2\omega^2)(1 - k^2\omega_2^2\omega^2)} = 3.8353 \frac{(0.27815 - \omega^2)(0.93731 - \omega^2)}{(1 - 0.22988\omega^2)(1 - 0.77464\omega^2)}$$

Problem 3.43 *Suppose that we wish to design a low-pass filter having an elliptic transducer power-gain characteristic. The peak-to-peak ripple within the passband, which extends from 0 to 100 MHz, must not exceed 1 dB, and at 120 MHz the gain must be attenuated by at least 30 dB. Determine the transducer power-gain characteristic.*

SOLUTION:

$$k = \frac{\omega_c}{\omega_s} = \frac{2\pi f_c}{2\pi f_s} = \frac{5}{6} = 0.83333$$

$$k' = \sqrt{1 - k^2} = 0.55277$$

$$K = F\left(k, \frac{\pi}{2}\right) = 2.067255, \quad K' = F\left(k', \frac{\pi}{2}\right) = 1.717152$$

$$\varepsilon = \sqrt{10^{0.1R_{dB}} - 1} = 0.508847$$

$$k_1 = \frac{\varepsilon}{\sqrt{10^{0.1A_{min}} - 1}} = \frac{0.508847}{\sqrt{10^3 - 1}} = 0.016099$$

$$k_1' = \sqrt{1 - k_1^2} = 0.99987$$

$$K_1 = F\left(k_1, \frac{\pi}{2}\right) = 1.570897, \quad K_1' = F\left(k_1', \frac{\pi}{2}\right) = 5.514046$$

$$n = \frac{K K_1'}{K' K_1} = 4.22579 \quad \rightarrow \quad n = 5$$

$$\omega_1 = \text{sn}\left(\frac{2K}{n}, k\right) = \text{sn}\,(0.826902, 0.83333) = \sin 44.15° = 0.69654$$

$$\omega_2 = \text{sn}\left(\frac{4K}{n}, k\right) = \text{sn}\,(1.653804, 0.83333) = \sin 76.65° = 0.97298$$

$$k_1 = k^5\left[\frac{(1-\omega_1^2)(1-\omega_2^2)}{(1-k^2\omega_1^2)(1-k^2\omega_2^2)}\right]^2 = 0.40188\left[\frac{(1-0.48517)(1-0.94669)}{(1-0.33692)(1-0.05742)}\right]^2$$

$$= 5.86762\times10^{-3}$$

$$A_{\text{dB}} = 10\log\left(1+\frac{\varepsilon^2}{k_1^2}\right) = 38.76 \text{ dB} > 30 \text{ dB}$$

$$H_0 = \sqrt{\frac{k^5}{k_1}} = 8.27591$$

$$F_5(\omega) = H_0\frac{\omega(\omega_1^2-\omega^2)(\omega_2^2-\omega^2)}{(1-k^2\omega_1^2\omega^2)(1-k^2\omega_2^2\omega^2)} = \frac{8.27591\omega(0.48517-\omega^2)(0.94669-\omega^2)}{(1-0.33692\omega^2)(1-0.65742\omega^2)}$$

The desired transducer power-gain characteristic is found to be

$$G(\omega^2) = \frac{H_5}{1+\varepsilon^2 F_5^2\left(\frac{\omega}{\omega_c}\right)}$$

where

$$\varepsilon = 0.508847, \quad \omega_c = 2\pi 10^8 \text{ rad/s}$$

Problem 3.44 *In Problem 3.43, suppose that the attenuation at 120 MHz is increased to at least 50 dB. Determine the transducer power-gain characteristic.*

SOLUTION:

$$k = \frac{\omega_c}{\omega_s} = \frac{5}{6} = 0.83333$$

$$k' = \sqrt{1 - k^2} = 0.55277$$

$$K = 2.067255, \quad K' = 1.717152$$

$$\varepsilon = \sqrt{10^{0.1 R_{dB}} - 1} = 0.508847$$

$$k_1 = \frac{\varepsilon}{\sqrt{10^{0.1 A_{min}} - 1}} = \frac{0.508847}{\sqrt{10^5 - 1}} = 1.609124 \times 10^{-3}$$

$$k_1' = \sqrt{1 - k_1^2} = 0.9999987$$

$$K_1 = F\left(k_1, \frac{\pi}{2}\right) = 1.570796, \quad K_1' = F\left(k_1', \frac{\pi}{2}\right) = 7.734817$$

$$n = \frac{K K_1'}{K' K_1} = 5.9281 \quad , \quad n = 6$$

$$\omega_1 = sn\left(\frac{K}{n}, k\right) = sn\,(0.34454, 0.83333) = \sin 19.48° = 0.33348$$

$$\omega_2 = sn\left(\frac{3K}{n}, k\right) = sn\,(1.03363, 0.83333) = \sin 53.37° = 0.80251$$

$$\omega_3 = sn\left(\frac{5K}{n}, k\right) = sn\,(1.72271, 0.83333) = \sin 78.94° = 0.98143$$

$$k_1 = k^6 \left[\frac{(1 - \omega_1^2)(1 - \omega_2^2)(1 - \omega_3^2)}{(1 - k^2\omega_1^2)(1 - k^2\omega_2^2)(1 - k^2\omega_3^2)}\right]^2$$

$$= 0.334898 \times \left[\frac{(1 - 0.11121)(1 - 0.64401)(1 - 0.96320)}{(1 - 0.07723)(1 - 0.44723)(1 - 0.66889)}\right]^2 = 1.59177 \times 10^{-3}$$

$$A_{dB} = 10 \log\left(1 + \frac{\varepsilon^2}{k_1^2}\right) = 50.0942 > 50 \text{ dB}$$

$$H_e = \sqrt{\frac{k^6}{k_1}} = 14.5049$$

$$F_6(\omega) = H_e \frac{(\omega_1^2 - \omega^2)(\omega_2^2 - \omega^2)(\omega_3^2 - \omega^2)}{(1 - k^2\omega_1^2\omega^2)(1 - k^2\omega_2^2\omega^2)(1 - k^2\omega_3^2\omega^2)}$$

$$= \frac{14.5049\,(0.11121 - \omega^2)(0.64401 - \omega^2)(0.96320 - \omega^2)}{(1 - 0.07723\omega^2)(1 - 0.44723\omega^2)(1 - 0.66889\omega^2)}$$

The desired transducer power-gain characteristic is obtained as

$$G(\omega^2) = \frac{H_6}{1 + \varepsilon^2 F_6^2\left(\frac{\omega}{\omega_c}\right)}$$

where

$$\varepsilon = 0.508847, \qquad \omega_c = 2\pi10^8 \text{ rad/s}$$

Problem 3.46 *Consider the transducer power-gain characteristic obtained in Problem 3.43. Determine the maximum and minimum values of gain in both the passband and the stopband and the frequencies at which these maxima and minima occur.*

SOLUTION: In the passband, we have

$$G\left(\omega^2\right)_{max} = H_5$$

at $\omega_{20}, \omega_{21}, \omega_{22}$, where

$$\omega_{20} = \omega_c \text{sn}(0, k) = 0$$

$$\omega_{21} = \omega_c \text{sn}\left(\frac{2K}{n}, k\right) = 2\pi10^8 \times 0.69654 = 4.3765\times10^8$$

$$\omega_{22} = \omega_c \text{sn}\left(\frac{4K}{n}, k\right) = 2\pi10^8 \times 0.97298 = 6.1134\times10^8$$

$$G\left(\omega^2\right)_{min} = \frac{H_5}{1+\epsilon^2} = 0.794328H_5$$

at ω'_1 and ω'_2, where

$$\omega'_1 = \omega_c \, \text{sn}\left(\frac{K}{n}, \, k\right) = \omega_c \, \text{sn} \, (0.413451, 0.83333) = 2.4792 \times 10^8$$

$$\omega'_2 = \omega_c \, \text{sn}\left(\frac{3K}{n}, \, k\right) = \omega_c \, \text{sn} \, (1,240353, 0.83333) = 5.5364 \times 10^8$$

In the stopband, we have

$$G\left(\omega^2\right)_{min} = 0$$

at ω_{p0}, ω_{p1}, and ω_{p2}, where

$$\omega_{p0} = \frac{\omega_c^2}{k\omega_{20}} = \infty$$

$$\omega_{p1} = \frac{\omega_e^2}{k\omega_{21}} = 10.8247 \times 10^0$$

$$\omega_{p2} = \frac{\omega_c^2}{k\omega_{22}} = 7.7492 \times 10^8$$

$$G\left(\omega^2\right)_{max} = \frac{H_5}{1+\dfrac{\epsilon^2}{k_1^2}} = 1.3295 \times 10^{-4} H_5$$

at ω''_1 and ω''_2, where

$$\omega''_1 = \frac{\omega_c^2}{k\omega'_1} = 19.1083 \times 10^8, \qquad \omega''_2 = \frac{\omega_c^2}{k\omega'_2} = 8.5569 \times 10^8$$

Problem 3.47 *Repeat Problem 3.46 for the transducer power-gain characteristic obtained in Problem 3.44.*

SOLUTION: In the passband, we have

$$G\left(\omega^2\right)_{max} = H_6$$

at ω_{21}, ω_{22} and ω_{23}, where

$$\omega_{21} = \omega_c \, sn\left(\frac{K}{n}, \, k\right) = 2\pi 10^8 \times 0.33348 = 2.0953 \times 10^8$$

$$\omega_{22} = \omega_c \, sn\left(\frac{3K}{n}, \, k\right) = 2\pi 10^8 \times 0.80251 = 5.0423 \times 10^8$$

$$\omega_{23} = \omega_c \, sn\left(\frac{5K}{n}, \, k\right) = 2\pi 10^8 \times 0.98143 = 6.1665 \times 10^8$$

$$G\left(\omega^2\right)_{min} = \frac{H_6}{1+\varepsilon^2} = 0.794328 H_6$$

at ω'_0, ω'_1 and ω'_2, where

$$\omega'_0 = \omega_c \, sn\,(0, k) = 0$$

$$\omega'_1 = \omega_c \, sn\left(\frac{2K}{n}, \, k\right) = \omega_c \, sn\,(0.689085, 0.83333) = 3.8276 \times 10^8$$

$$\omega'_2 = \omega_c \, sn\left(\frac{4K}{n}, \, k\right) = \omega_c \, sn\,(1.37817, 0.83333) = 5.7833 \times 10^8$$

In the stopband, we have

$$G\left(\omega^2\right)_{min} = 0$$

at ω_{p1}, ω_{p2} and ω_{p3}, where

$$\omega_{p1} = \frac{\omega_c^2}{k\omega_{21}} = 22.6095 \times 10^8$$

$$\omega_{p2} = \frac{\omega_c^2}{k\omega_{22}} = 9.3953\times10^8$$

$$\omega_{p3} = \frac{\omega_c^2}{k\omega_{23}} = 7.6825\times10^8$$

$$G\left(\omega^2\right)_{max} = \frac{H_6}{1 + \left|\frac{\varepsilon}{k_1}\right|^2} = 9.785\times10^{-6}H_6$$

at ω''_0, ω''_1 and ω''_2, where

$$\omega''_0 = \frac{\omega_c^2}{k\omega'_0} = \infty$$

$$\omega''_1 = \frac{\omega_c^2}{k\omega'_1} = 12.3771\times10^8$$

$$\omega''_2 = \frac{\omega_c^2}{k\omega'_2} = 8.1916\times10^8$$

Problem 3.51 *Using the specifications stated in Problem 3.43, compute the input impedance function* $Z_{11}(s)$ *of the filter.*

SOLUTION: Given

$$n=5, \quad k = 5/6 = 0.83333, \quad k' = 0.55277, \quad K = 2.067255, \quad \varepsilon = 0.508847$$

$$k_1 = 5.86762\times10^{-3}, \quad K_1 = K(k_1) = F\left(k_1, \frac{\pi}{2}\right) = 1.570811$$

$$a = \frac{K}{nK_1}F\left[\sin^{-1}\left(1 + \varepsilon^2\right)^{-1/2}, \quad k'_1\right] = 0.26321F(63.03°, 0.99998)$$

$$= 0.26321 \times 1.42792 = 0.37584$$

where

$$k'_1 = \sqrt{1 - k_1^2} = 0.99998$$

$$y_{p0} = j \, \text{sn} \, (ja, k) = - \, \text{tn} \, (a, k') = - \, \text{tn} \, 21.38° = - 0.391493$$

$$y_{p1} = j \, \text{sn} \left(\pm \frac{2K}{n} + ja, k \right) = j \, \text{sn} \, (\pm 0.8269 + j0.37584, 0.83333)$$

$$= - 0.2175069 \pm j0.748181$$

$$y_{p2} = j \, \text{sn} \left(\pm \frac{4K}{n} + ja, k \right) = j \, \text{sn} \, (\pm 1.653804 + j0.37584, 0.83333)$$

$$= - 0.4.80655 \times 10^{-2} \pm j0.998477$$

where

$$\text{sn} \, (0.826902, k) = \sin 44.15°$$

$$\text{sn} \, (a, k') = \sin 21.38°$$

$$\text{sn} \, (1.653804, k) = \sin 76.65°$$

giving

$$r(y) = (y - y_{p0})(y - y_{p1})(y - \bar{y}_{p1})(y - y_{p2})(y - \bar{y}_{p2})$$

$$= y^5 + a_4 y^4 + a_3 y^3 + a_2 y^2 + a_1 y + a_0$$

where

$$a_4 = 0.9226378, \quad a_3 = 1.856108, \quad a_2 = 1.138301$$

$$a_1 = 0.799666, \quad a_0 = 0.2374948$$

Let $H_5 = 1$. $\hat{r}(y)$ is formed by the simple zeros of $F_n^2(-jy)$.

$$\omega_1 = \text{sn} \, (2K/n, k) = 0.69654, \qquad \omega_2 = \text{sn} \, (4K/n, k) = 0.97298$$

$$\hat{r}(y) = y(y^2 + \omega_1^2)(y^2 + \omega_2^2) = y(y^2 + 0.69654^2)(y^2 + 0.97298^2)$$

$$= y^5 + b_4 y^4 + b_3 y^3 + b_2 y^2 + b_1 y + b_0$$

where

$$b_4 = 0, \quad b_3 = 1.431858, \quad b_2 = 0, \quad b_1 = 0.4593, \quad b_0 = 0$$

$$S_{11}(s) = \pm \frac{\hat{r}(y)}{r(y)}$$

The input impedance $Z_{11}(s)$ is obtained as

$$\frac{Z_{11}}{R_1} = \frac{r(y) + \hat{r}(y)}{r(y) - \hat{r}(y)} = \frac{2y^5 - 0.9226y^4 + 3.288y^3 + 1.1383y^2 + 1.259y + 0.2375}{0.9226y^4 + 0.4242y^3 + 1.1383y^2 + 0.3404y + 0.2375}$$

Problem 3.52 *Using the specifications stated in Problem 3.44, compute the input impedance function $Z_{11}(s)$ of the filter.*

SOLUTION: Given

$$n = 6, \quad k = 5/6 - 0.83333, \quad k' = 0.55277, \quad K = 2.067255, \quad \varepsilon = 0.508847$$

$$k_1 = 1.59177 \times 10^{-3}, \quad K_1 = K(k_1) = F\left(k_1, \frac{\pi}{2}\right) = 1.570796$$

$$a = \frac{K}{nK_1} F\left[\sin^{-1}\left(1 + \varepsilon^2\right)^{-1/2}, \ k'_1\right] = 0.21934F(63.03°, \ 0.9999987)$$

$$= 0.21934 \times 1.42794 = 0.3132181$$

where

$$k'_1 = \sqrt{1 - k_1^2} = 0.9999987$$

obtaining

$$y_{p1} = j \operatorname{sn}\left(\pm \frac{K}{n} + ja, \ k\right) = j \operatorname{sn}\left(\pm 0.3445425 + ja, \ k\right) = -0.289489 \pm j0.359887$$

where

$$sn\,(0.3445425,\,k)\;=\;\sin 19.48°, \qquad sn\,(a,\,k')\;=\;\sin 17.86°$$

$$y_{p2}\;=\;j\,sn\left(\pm\frac{3K}{n}+ja,\,k\right)\;=\;j\,sn\,(\pm\,1.0336275+ja,\,k)\;=\;-\,0.136593\pm j0.834265$$

where

$$sn\,(1.0336275,\,k)\;=\;\sin 53.37°$$

$$y_{p3}\;=\;j\,sn\left(\pm\frac{5K}{n}+ja,\,k\right)\;=\;j\,sn\,(\pm\,1.72271+ja,\,k)\;=\;-\,3.325929\times10^{-2}\pm j0.998312$$

where

$$sn\,(1.72271,\,k)\;=\;\sin 78.94°$$

Finally, we have

$$r(y)\;=\;(y-y_{p1})(y-\bar{y}_{p1})(y-y_{p2})(y-\bar{y}_{p2})(y-y_{p3})(y-\bar{y}_{p3})$$

$$=\;y^6+a_5y^5+a_4y^4+a_3y^3+a_2y^2+a_1y+a_0$$

where

$$a_5\;=\;0.9186826,\qquad a_4\;=\;2.140566,\qquad a_3\;=\;1.394529$$

$$a_2\;=\;1.267538,\qquad a_1\;=\;0.481118,\qquad a_0\;=\;0.1522107$$

Let

$$R_2\;=\;2R_1,\qquad H_6\;=\;0.9814$$

$$\hat{\varepsilon}\;=\;\varepsilon\left(1-H_6\right)^{-1/2}\;=\;3.7310$$

obtaining

$$\hat{y}_{p1}=-\,0.05268\pm j0.3343,\quad \hat{y}_{p2}=-\,0.02577\pm j0.8036,\quad \hat{y}_{p3}=-\,0.006408\pm j0.9820$$

$$\hat{r}(y)\;=\;y^6+0.1697y^5+1.7328y^4+0.2352y^3+0.8141y^2+0.07233y+0.07142$$

$$\lambda\;=\;\left[1-\frac{H_6}{1+\left(\dfrac{\varepsilon}{k_1}\right)^2}\right]^{\frac{1}{2}}\;=\;\sqrt{1-\frac{0.9814}{1+(0.508847/0.001592)^2}}\;=\;0.999995\approx1$$

38

$$S_{11}(s) = \pm \frac{y^6 + 0.1697y^5 + 1.7328y^4 + 0.2352y^3 + 0.8141y^2 + 0.07233y + 0.071}{y^6 + 0.9187y^5 + 2.1406y^4 + 1.3945y^3 + 1.2675y^2 + 0.4811y + 0.1521}$$

$$\frac{Z_{11}(s)}{R_1} = \frac{r(y) + \lambda \hat{r}(y)}{r(y) - \lambda \hat{r}(y)}$$

$$= \frac{2y^6 + 1.0884y^5 + 3.8734y^4 + 1.6297y^3 + 2.0816y^2 + 0.5534y + 0.2235}{0.749y^5 + 0.4078y^4 + 1.1593y^3 + 0.4534y^2 + 0.4288y + 0.080}$$

Problem 3.58 *Suppose that we wish to design a filter having a transducer power-gain characteristic that gives at most 1-dB passband ripple and at least 30-dB attenuation in the stopband at 1.1 times the cutoff frequency. Determine the values of n for the Butterworth, Chebyshev and elliptic responses. Also compare your results.*

SOLUTION:

(1) For the Butterworth response, we have

$$n \geq \frac{\log(10^3 - 1) - \log(10^{0.1} - 1)}{2 \log 1.1} = 43.3 \quad \rightarrow \quad n = 44$$

(2) For the Chebyshev response, we have

$$\varepsilon = \sqrt{10^{0.1 \times 1} - 1} = 0.508847$$

$$n \geq \frac{\frac{1}{2} \ln\left(4 \times 10^3 / 0.508847\right)}{\ln\left(1.1 + \sqrt{1.1^2 - 1}\right)} = 10.11 \quad \rightarrow \quad n = 11$$

(3) For the elliptic response, we have

$$k = \frac{\omega_c}{\omega_s} = \frac{1}{1.1} = 0.9090909, \qquad k' = \sqrt{1 - k^2} = 0.416598$$

$$\varepsilon = \sqrt{10^{0.1} - 1} = 0.508847$$

$$k_1 = \frac{\varepsilon}{\sqrt{10^3 - 1}} = 0.016099, \qquad k'_1 = \sqrt{1 - k_1^2} = 0.99987$$

$$K = F\left(k, \frac{\pi}{2}\right) = 2.321924, \qquad K' = F\left(k, \frac{\pi}{2}\right) = 1.64653$$

$$K_1 = F\left(k_1, \frac{\pi}{2}\right) = 1.570897, \qquad K'_1 = F\left(k'_1, \frac{\pi}{2}\right) = 5.515653$$

$$n = \frac{KK'_1}{K'K_1} = 4.95 \quad \rightarrow \quad n = 5$$

showing that the order of the elliptic response is lower than others.

Problem 3.59 *Compute the value* sn^{-1} *(j/2.36571, 0.007918) and compare your result with (3.258a).*

SOLUTION:

$$sn^{-1}(j/2.36571, \ 0.007918) = sn^{-1}(j/\varepsilon, k_1)$$

$$= jF\left[\sin^{-1}\left(1 + \varepsilon^2\right)^{-1/2}, \ k'_1\right] = jF(22.914°, \ 0.99996865) = j0.4110332$$

confirming the result in (3.258a).

Problem 3.63 *Design a high-pass filter whose radian cutoff frequency is 10^6 rad/s. The filter is required to have an equiripple transducer power-gain characteristic and is to be operated between a resistive generator of internal resistance 50 Ω and a 150 Ω load. The peak-to-peak ripple in the passband must not exceed 1.5 dB and at 3×10^5 rad/s or less the gain must be at least 50 dB down from its peak value in the passband. Also plot the gain response versus ω.*

SOLUTION: The corresponding normalized low-pass network has the following specifications:

$$\omega_c = 1, \qquad \omega_s = \frac{10}{3} = 3.3333, \qquad \frac{\omega_s}{\omega_c} = 3.3333 = k$$

$$\varepsilon = \sqrt{10^{0.1 \times 1.5} - 1} = 0.64229$$

$$n \geq \frac{\frac{1}{2} \ln (4 \times 10^5 / \varepsilon)}{\ln \left(k + \sqrt{k^2 - 1}\right)} = 3.56 \quad \rightarrow \quad n = 4$$

$$G_{min} = 1 - \left(\frac{R_2 - R_1}{R_2 + R_1}\right)^2 = 1 - \left(\frac{100}{200}\right)^2 = \frac{3}{4}$$

$$K_4 = G_{min}\left(1 + \varepsilon^2\right) = 1.0594 > 1$$

Let $K_4 = 1$. Then

$$\varepsilon = \sqrt{\frac{K_4}{G_{min}} - 1} = \frac{1}{\sqrt{3}} = 0.57735$$

$$\sinh a = 0.33522, \qquad \sinh \hat{a} = 0$$

yielding

$$L_1 = 114.1588 \text{ H}, \qquad C_2 = 2.022973 \times 10^{-2} \text{ F}$$
$$L_3 = 151.7227 \text{ H}, \qquad C_4 = 1.522124 \times 10^{-2} \text{ F}$$

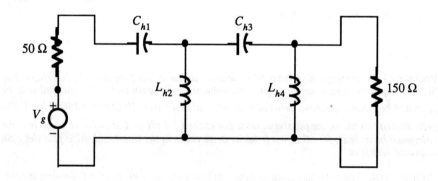

The element values of the desired Chebyshev high-pass filter are found to be

$$C_{h1} = \frac{1}{L_1 \omega_c} = 8.75973 \text{ nF}, \qquad L_{h2} = \frac{1}{C_2 \omega_c} = 49.4322 \text{ μH}$$

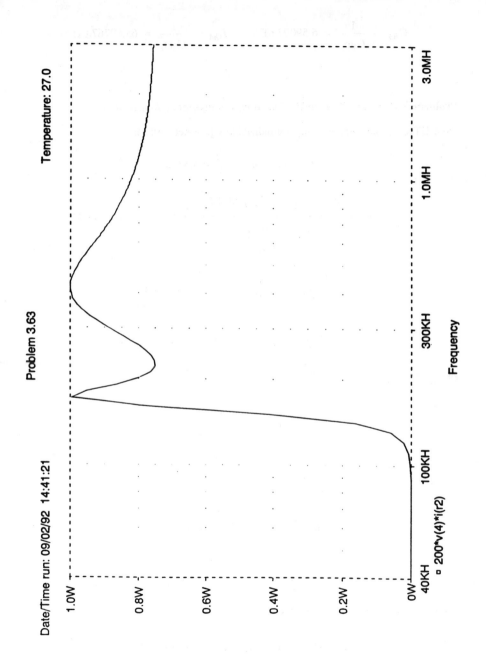

Problem 3.63

Date/Time run: 09/02/92 14:41:21

Temperature: 27.0

□ 200*v(4)*i(r2)

Frequency

$$C_{h3} = \frac{1}{L_3 \omega_c} = 6.59097 \text{ nF}, \qquad L_{h4} = \frac{1}{C_4 \omega_c} = 65.69767 \text{ μH}$$

Problem 3.64 *Repeat Problem 3.63 for a maximally-flat high-pass filter.*

SOLUTION: The corresponding normalized low-pass network has

$$\omega_c = 1, \qquad \omega_s = \frac{10}{3} = 3.3333$$

$$n \geq \frac{\log(10^{0.1 \times 50} - 1) - \log(10^{0.1 \times 1.5} - 1)}{2 \log\left(\dfrac{10^6}{3 \times 10^5}\right)} = 5.15 \quad \rightarrow \quad n = 6$$

$$\delta^6 = \frac{R_2 - R_1}{R_2 + R_1} = \frac{100}{200} = \frac{1}{2} \quad \rightarrow \quad \delta = 0.890899$$

$$K_6 = 1 - \left(\delta^6\right)^2 = \frac{3}{4} = 0.75$$

Let $\omega_{cl} = 1$. From

$$G\left(\omega^2\right) = \frac{K_6}{1 + \left(\dfrac{\omega_l}{\omega_{cl}}\right)^{2n}}$$

we have

$$\omega_{sl} = \left(10^{0.1 \times 50} - 1\right)^{1/12} = 2.610155$$

$$\omega_{rl} = \left(10^{0.1 \times 1.5} - 1\right)^{1/12} = 0.928871$$

From the high-pass response, we obtain

$$\omega_{rc} = \omega_r \omega_{rl} = 10^6 \times 0.928871 = 9.28871 \times 10^5$$

$$\omega_{sc} = \omega_s \omega_{sl} = 3 \times 10^5 \times 2.610155 = 7.830465 \times 10^5$$

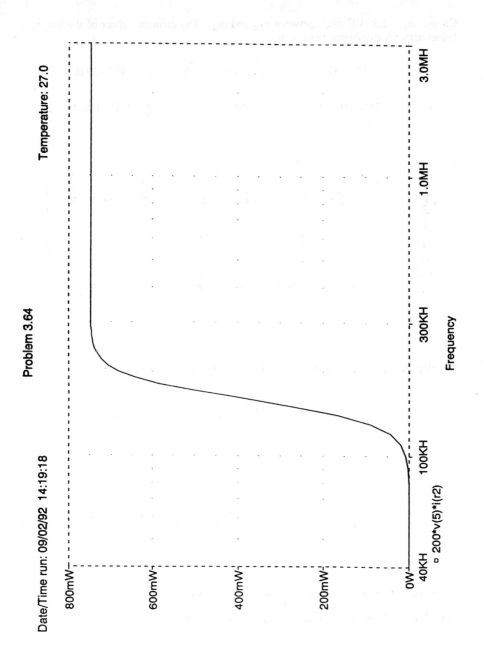

Date/Time run: 09/02/92 14:19:18

Problem 3.64

Temperature: 27.0

□ 200*v(5)*i(r2)

Frequency

44

Choose $\omega_c = 8.5 \times 10^5$ rad/s between ω_{rc} and ω_{sc}. The element values of the low-pass ladder network are obtained as follows:

$$L_{l1} = 237.2287 \text{ H}, \quad C_{l2} = 1.231291 \times 10^{-2} \text{F}, \quad L_{l3} = 245.7739 \text{ H}$$

$$C_{l4} = 8.465688 \times 10^{-3} \text{ F}, \quad L_{l5} = 120.2121 \text{ H}, \quad C_{l6} = 1.825028 \times 10^{-3} \text{ F}$$

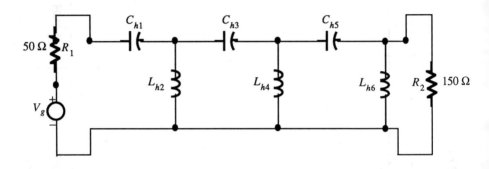

The element values of the desired maximally-flat high-pass filter are

$$C_{h1} = \frac{1}{L_{l1}\omega_c} = 4.959225 \text{ nF}, \quad L_{h2} = \frac{1}{C_{l2}\omega_c} = 95.547729 \text{ μH}$$

$$C_{h3} = \frac{1}{L_{l3}\omega_c} = 4.786800 \text{ nF}, \quad L_{h4} = \frac{1}{C_{l4}\omega_c} = 138.969238 \text{ μH}$$

$$C_{h5} = \frac{1}{L_{l5}\omega_c} = 9.786624 \text{ nF}, \quad L_{h6} = \frac{1}{C_{l6}\omega_c} = 644.631528 \text{ μH}$$

Problem 3.66 *Repeat Problem 3.63 for an elliptic high-pass filter.*

SOLUTION: The corresponding normalized low-pass network has

$$\omega_c = 1, \quad \omega_s = \frac{10}{3} = 3.3333$$

$$k = \frac{\omega_c}{\omega_s} = \frac{3}{10} = 0.3, \qquad K = F\left(k, \frac{\pi}{2}\right) = 1.608049$$

$$k' = \sqrt{1 - k^2} = 0.953939, \qquad K' = F\left(k', \frac{\pi}{2}\right) = 2.627772$$

$$\varepsilon = \sqrt{10^{0.15} - 1} = 0.642291$$

$$k_1 = \frac{\varepsilon}{\sqrt{10^5 - 1}} = 2.031112 \times 10^{-3}, \qquad K_1 = F\left(k_1, \frac{\pi}{2}\right) = 1.570796$$

$$k_1' = \sqrt{1 - k_1^2} = 0.9999979, \qquad K_1' = F\left(k_1', \frac{\pi}{2}\right) = 7.509537$$

$$n = \frac{K K_1'}{K' K_1} = 2.93 \quad \rightarrow \quad n = 3$$

Given $n = 3$, $\varepsilon = 0.642291$ and $k = 0.3$, we obtain

$$k_1 = 1.810969 \times 10^{-3}$$

$$A_{dB} = 10 \log\left(1 + \frac{\varepsilon^2}{k_1^2}\right) = 50.996 \text{ dB} > 50 \text{ dB}$$

meeting the requirement.

$$K = 1.6080, \qquad K_1 = 1.5708$$

The poles are found to be

$$y_0 = -0.432, \qquad y_{12} = -0.2023 \pm j0.9447$$

obtaining

$$r(y) = y^3 + 0.8365y^2 + 1.1082y + 0.4032$$

$$\hat{r}(y) = y(y^2 + 0.7587), \qquad S_{11}(y) = \pm \frac{\hat{r}(y)}{r(y)}$$

$$\frac{Z_{11}(y)}{R_1} = \frac{2y^3 + 0.8365y^2 + 1.8669y + 0.4032}{0.8365y^2 + 0.3495y + 0.4032}$$

Problem 3.67 *Design a band-pass filter whose passband is from 10^6 rad/s to 5×10^6 rad/s. The filter is required to have an equiripple transducer power-gain characteristic and is to be operated between a resistive generator of internal resistance 50 Ω and a 150-Ω load. The peak-to-peak ripple in the passband must not exceed 1-dB and at $\omega = 15\times10^6$ rad/s the gain must be at least 60 dB down from its peak value in the passband. Also plot the gain response versus ω.*

SOLUTION:

$$B = \omega_2 - \omega_1 = 4\times10^6 \text{ rad/s}$$

$$\omega_0 = \sqrt{\omega_1\omega_2} = \sqrt{5}\times10^6 = 2.236067977\times10^6 \text{ rad/s}$$

$$\omega'_s = \frac{\omega_0}{B}\left(\frac{\omega_s}{\omega_0} - \frac{\omega_0}{\omega_s}\right) = \frac{\sqrt{5}\times10^6}{4\times10^6}\left(\frac{15\times10^6}{\sqrt{5}\times10^6} - \frac{\sqrt{5}\times10^6}{15\times10^6}\right)$$

$$= \frac{1}{4}\left(15 - \frac{1}{3}\right) = \frac{44}{12} = \frac{11}{3} = 3.66667$$

$$\omega'_c = 1 \text{ rad/s}, \qquad k = \frac{\omega'_s}{\omega'_c} = \frac{11}{3}$$

$$\varepsilon = \sqrt{10^{0.1} - 1} = 0.508847$$

$$n \geq \frac{\frac{1}{2}\ln(4\times10^6/\varepsilon)}{\ln(k + \sqrt{k^2 - 1})} = 4.023 \quad \rightarrow \quad n = 5$$

$$K_5 = 1 - \left(\frac{R_2 - R_1}{R_2 + R_1}\right)^2 = 1 - \left(\frac{100}{200}\right)^2 = \frac{3}{4} = 0.75$$

$$\sinh a = 0.2894934, \qquad \sinh \hat{a} = 0$$

$$L_1 = 106.744 \text{ H}, \qquad C_2 = 2.182216\times10^{-2}\text{F}, \qquad L_3 = 150.0461 \text{ H}$$

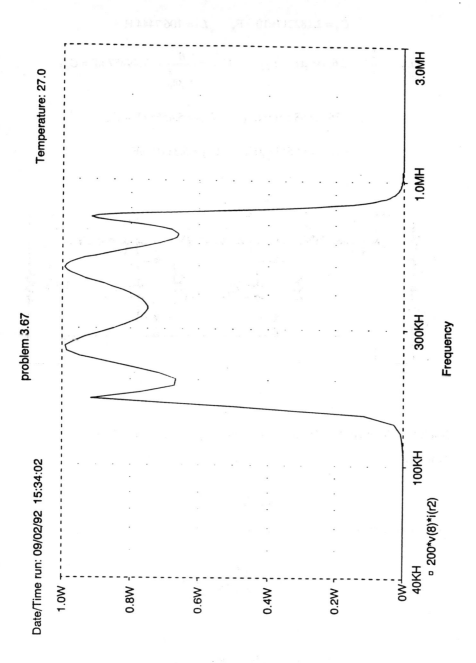

$$C_4 = 2.182217 \times 10^{-2} \, \text{F}, \qquad L_5 = 106.7444 \, \text{H}$$

$$L_{b1} = \frac{L_1}{B} = 26.686 \, \mu\text{H} = L_{b5}, \qquad C_{b1} = \frac{B}{L_1 \omega_0^2} = 7.49457 \, \text{nF} = C_{b5}$$

$$L_{b2} = 36.65998 \, \mu\text{H} = L_{b4}, \qquad C_{b2} = 5.4555 \, \text{nF} = C_{b4}$$

$$L_{b3} = 37.5115 \, \mu\text{H}, \qquad C_{b2} = 5.33169 \, \text{nF}$$

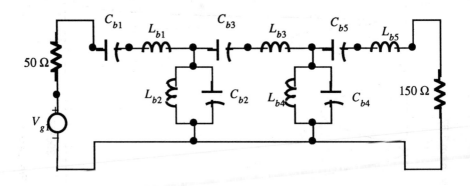

Problem 3.68 *Repeat Problem 3.67 for a maximally-flat band-pass filter.*

SOLUTION:

$$\omega_0^2 = \omega_1 \omega_2 = 5 \times 10^6$$

and the bandwidth B is unknown. Since

$$\omega_l = \frac{1}{B}\left(\omega_B - \frac{\omega_0^2}{\omega_B}\right)$$

we obtain

$$\omega_{rl} = \frac{1}{B}(\omega_2 - \omega_1) = \frac{4 \times 10^6}{B}$$

$$\omega_{sl} = \frac{1}{B}\left(15\times10^6 - \frac{5\times10^{12}}{15\times10^6}\right) = \frac{14.666667\times10^6}{B}$$

$$n \geq \frac{\log\left(10^{0.1\times60} - 1\right) - \log\left(10^{0.1\times1} - 1\right)}{2\log\dfrac{14.666667\times10^6}{4\times10^6}} = 5.837 \qquad \rightarrow \qquad n = 6$$

Choose $n = 6$. Then

$$K_6 = 1 - \left(\frac{R_2 - R_1}{R_2 + R_1}\right)^2 = \frac{3}{4} = 0.75$$

$$\delta^6 = \frac{R_2 - R_1}{R_2 + R_1} = \frac{100}{200} = 0.5, \qquad \delta = (0.5)^{1/6} = 0.890899$$

The corresponding normalized Butterworth low-pass filter has the following specifications:

$$G\!\left(\omega^2\right) = \frac{K_6}{1 + \left(\dfrac{\omega_l}{\omega_c}\right)^{12}}$$

$$\omega_{rl} = \left(10^{0.1\times1} - 1\right)^{1/12} = 0.893507 = \frac{4\times10^6}{B_r} \qquad \rightarrow \qquad B_r = 4.476742\times10^6$$

$$\omega_{sl} = \left(10^{0.1\times60} - 1\right)^{1/12} = 3.162277397 = \frac{14.666667\times10^6}{B_s}$$

$$\rightarrow \qquad B_s = 4.638007726\times10^6$$

Choose $B = 4.6\times10^6$ between B_r and B_s. The element values of the low-pass ladder network are given as follows:

$$L_{l1} = 237.2287 \text{ H}, \qquad C_{l2} = 1.231291\times10^{-2} \text{ F}$$

$$L_{l3} = 245.7739 \text{ H}, \qquad C_{l4} = 8.465688 \times 10^{-3} \text{ F}$$

$$L_{b5} = 120.2121 \text{ H}, \qquad C_{b5} = 1.825028 \times 10^{-3} \text{ F}$$

The desired Butterworth band-pass filter is shown below:

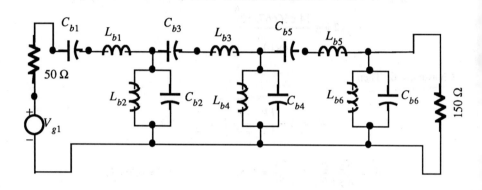

$$L_{b1} = \frac{L_{l1}}{B} = 51.571457 \text{ µH}, \qquad C_{b1} = \frac{B}{L_{l1}\omega_0^2} = \frac{1}{L_1\omega_0^2} = 3.878114 \text{ nF}$$

$$L_{b2} = \frac{B}{C_{l2}\omega_0^2} = \frac{1}{C_2\omega_0^2} = 74.718324 \text{ µH}, \qquad C_{b2} = \frac{C_{l2}}{B} = 2.676720 \text{ nF}$$

$$L_{b3} = \frac{L_{l3}}{B} = 53.429109 \text{ µH}, \qquad C_{b3} = \frac{1}{L_3\omega_0^2} = 3.743278 \text{ nF}$$

$$L_{b4} = \frac{1}{C_4\omega_0^2} = 0.108674 \text{ mH}, \qquad C_{b4} = \frac{C_{l4}}{B} = 1.840367 \text{ nF}$$

$$L_{b5} = \frac{L_{l5}}{B} = 26.133065 \text{ µH}, \qquad C_{b5} = \frac{1}{L_5\omega_0^2} = 7.653140 \text{ nF}$$

$$L_{b6} = \frac{1}{C_6\omega_0^2} = 0.504102 \text{ mH}, \qquad C_{b6} = \frac{C_{l6}}{B} = 0.396745 \text{ nF}$$

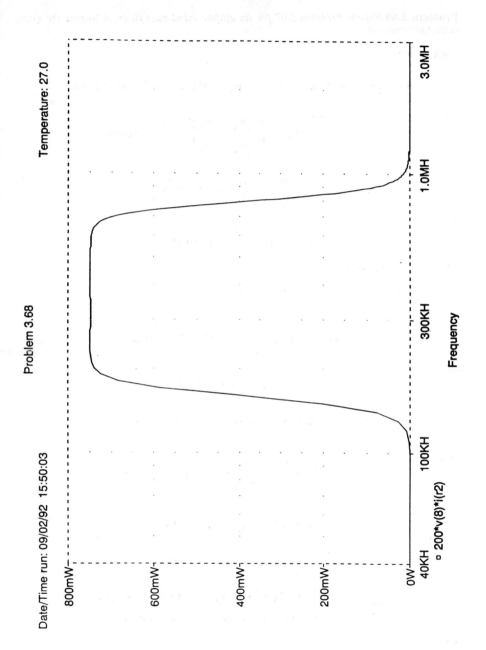

Problem 3.68

Date/Time run: 09/02/92 15:50:03

Temperature: 27.0

□ 200*v(8)*i(r2)

Frequency

Problem 3.69 *Repeat Problem 3.67 for an elliptic band-pass filter. Also plot the gain response versus ω.*

SOLUTION:

$$B = \omega_2 - \omega_1 = 4 \times 10^6 \, \text{rad/s}, \qquad \omega_0 = \sqrt{\omega_1 \omega_2} = 2.2361 \times 10^6 \, \text{rad/s}$$

$$\omega'_c = 1, \qquad \omega'_s = \frac{\omega_0}{B}\left(\frac{\omega_s}{\omega_0} - \frac{\omega_0}{\omega_s}\right) = \frac{11}{3} = 3.66667$$

$$k = \frac{\omega'_c}{\omega'_s} = \frac{3}{11} = 0.27272727$$

$$\varepsilon = \sqrt{10^{0.1} - 1} = 0.508847$$

Choose $n = 4$ and we obtain

$$k_1 = 9.3378 \times 10^{-5}$$

$$A_{dB} = 10 \log\left(1 + \frac{\varepsilon^2}{k_1^2}\right) = 74.73 \, \text{dB} > 60 \, \text{dB}$$

$$K = 1.6013, \qquad K_1 = 1.5708$$

$$r(y) = y^4 + 0.9508y^3 + 1.4616y^2 + 0.7518y + 0.2864$$

$$H_4 = \left(1 + \varepsilon^2\right)\left[1 - \left(\frac{R_2/R_1 - 1}{R_2/R_1 + 1}\right)^2\right] = 0.94419$$

$$\hat{\varepsilon} = \varepsilon\left(1 - H_4\right)^{-1/2} = 2.1540$$

$$\hat{r}(y) = y^4 + 0.2939y^3 + 1.0528y^2 + 0.1969y + 0.1432$$

$$\frac{Z_{11}(s)}{R_1} = \frac{2y^4 + 1.2447y^3 + 2.5144y^2 + 0.9487y + 0.4296}{0.6569y^3 + 0.4088y^2 + 0.5549y + 0.1432}$$

where

$$\lambda = \left[1 - \frac{0.94419}{1 + \left(\frac{0.50884}{9.3378\times10^{-5}}\right)^2}\right]^{1/2} = 0.99999998 \approx 1$$

Problem 3.70 *In Problem 3.67, suppose that from 10^6 rad/s to 5×10^6 rad/s is the rejection bandwidth. Design an equiripple band-elimination filter that gives at least 60 dB attenuation at 1.8×10^6 rad/s, everything else being the same.*

SOLUTION:

$$B = \omega_2 - \omega_1 = 4\times10^6 \text{ rad/s}, \qquad \omega_0^2 = 5\times10^{12}$$

The corresponding normalized low-pass specifications are given as follows:

$$\omega'_c = 1 \text{ rad/s}$$

$$\omega'_s = \frac{B}{\frac{\omega_0^2}{\omega_s} - \omega_s} = 4.0909091, \qquad K = \frac{\omega'_s}{\omega'_c} = 4.0909091$$

$$\varepsilon = \sqrt{10^{0.1} - 1} = 0.508847$$

$$n \geq \frac{\frac{1}{2}\ln\left(4\times10^6/\varepsilon\right)}{\ln\left(k + \sqrt{k^2 - 1}\right)} = 3.8 \quad \rightarrow \quad n = 4$$

$$G_{min} = 1 - \left(\frac{R_2 - R_1}{R_2 + R_1}\right)^2 = 1 - \frac{1}{4} = \frac{3}{4}$$

$$K_4 = G_{min}(1 + \varepsilon^2) = 0.944193952$$

$$L_1 = 104.9525 \text{ H}, \qquad C_2 = 2.128883\times10^{-2} \text{ F}$$

$$L_3 = 141.5558 \text{ H}, \qquad C_4 = 1.578404\times10^{-2} \text{ F}$$

54

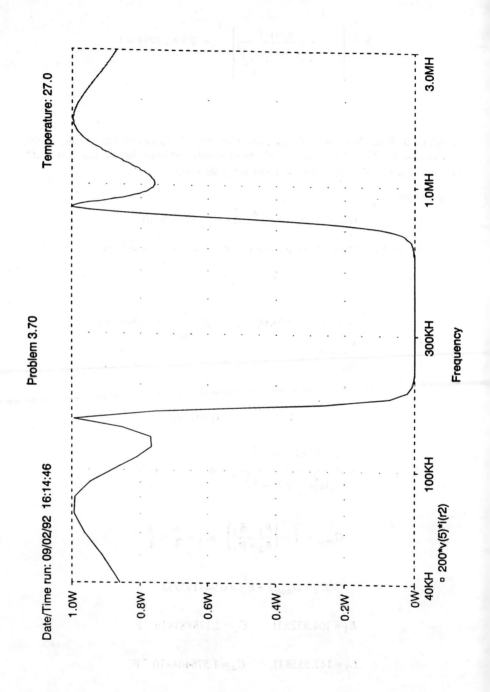

$$L_{e1} = \frac{L_1 B}{\omega_0^2} = 83.962 \ \mu H, \qquad C_{e1} = \frac{1}{L_1 B} = 2.38203 \ nF$$

$$L_{e2} = \frac{1}{C_2 B} = 11.74325 \ \mu H, \qquad C_{e2} = \frac{C_2 B}{\omega_0^2} = 17.03106 \ nF$$

$$L_{e3} = 0.11324 \ mH, \qquad C_{e3} = 1.76609 \ nF$$

$$L_{e4} = 15.83878 \ \mu H, \qquad C_{e4} = 12.62723 \ nF$$

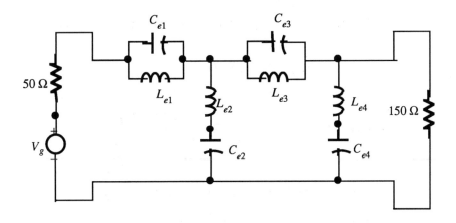

Problem 3.71 *Repeat Problem 3.70 for an elliptic band-elimination filter and plot its gain response versus ω.*

SOLUTION:

$$B = \omega_2 - \omega_1 = 4 \times 10^6 \ rad/s, \qquad \omega_0^2 = 5 \times 10^{12}$$

The corresponding normalized low-pass specifications are given by

$$\omega'_c = 1 \ rad/s, \qquad \omega'_s = \frac{B}{\frac{\omega_0^2}{\omega_s} - \omega_s} = 4.0909$$

$$\varepsilon = \sqrt{10^{0.1} - 1} = 0.508847, \qquad k = \frac{\omega'_c}{\omega'_s} = 0.24444$$

Choose $n = 4$ and we obtain

$$k_1 = 5.9361 \times 10^{-5}$$

$$A_{dB} = 10 \log\left(1 + \frac{\varepsilon^2}{k_1^2}\right) = 78.6 \, dB \; > \; 60 \, dB$$

$$K = 1.595084, \qquad K_1 = 1.570796$$

$$r(y) = y^4 + 0.9513y^3 + 1.4601y^2 + 0.75y + 0.2842$$

$$H_4 = \left(1 + \varepsilon^2\right)\left[1 - \left(\frac{R_2/R_1 - 1}{R_2/R_1 + 1}\right)^2\right] = 0.94419$$

$$\hat{\varepsilon} = \varepsilon(1 - H_4)^{-1/2} = 2.1540, \qquad \lambda = \sqrt{1 - \frac{H_4}{1 + \left(\frac{\varepsilon}{k_1}\right)^2}} \approx 1$$

$$\hat{r}(y) = y^4 + 0.2938y^3 + 1.0508y^2 + 0.1962y + 0.1421$$

$$\frac{Z_{11}(s)}{R_1} = \frac{2y^4 + 1.2451y^3 + 2.5109y^2 + 0.9462y + 0.4263}{0.6575y^3 + 0.4093y^2 + 0.5538y + 0.1421}$$

Problem 3.72 *Repeat Problem 3.70 for a maximally-flat band-elimination filter and plot its gain response versus ω.*

SOLUTION: The low-pass to band-elimination transformation is given by

$$\omega_l = \frac{B}{\dfrac{\omega_0^2}{\omega_e} - \omega_e}$$

where

$$\omega_0^2 = \omega_1 \omega_2 = 5 \times 10^{12}$$

and the bandwidth B is unknown. At ω_1 we have

$$\omega_{rl} = \frac{B}{\dfrac{5 \times 10^{12}}{10^6} - 10^6} = \frac{B}{4 \times 10^6}$$

and at ω_2

$$\omega_{sl} = \frac{B}{\dfrac{5 \times 10^{12}}{1.8 \times 10^6} - 1.8 \times 10^6} = \frac{B}{0.977778 \times 10^6}$$

giving

$$n \geq \frac{\log (10^{0.1 \times 60} - 1) - \log (10^{0.1 \times 1} - 1)}{2 \log \dfrac{4 \times 10^6}{0.977778 \times 10^6}} = 5.3829 \quad \rightarrow \quad n = 6$$

$$\delta = \sqrt[6]{\frac{R_2 - R_1}{R_2 + R_1}} = \sqrt[6]{\frac{100}{200}} = 0.890899$$

$$K_6 = 1 - \left(\delta^6\right)^2 = \frac{3}{4} = 0.75$$

The corresponding normalized low-pass response is given by

$$G\left(\omega^2\right) = \frac{K_6}{1 + \left(\dfrac{\omega_l}{\omega_c}\right)^2}$$

where $\omega_c = 1$.

$$\omega_{rl} = \left(10^{0.1 \times 1} - 1\right)^{\frac{1}{12}} = 0.893507 = \frac{B_r}{4 \times 10^6} \quad \rightarrow \quad B_r = 3.574028 \times 10^6$$

$$\omega_{sl} = \left(10^{0.1\times60} - 1\right)^{\frac{1}{12}} = 3.162277397 = \frac{B_s}{0.977778\times10^6} \rightarrow B_s = 3.0920045\times10^6$$

Choose

$$B = 3.3\times10^6 \text{ rad/s}$$

between B_r and B_s. The element values of the low-pass ladder network are found to be

$$L_{l1} = 237.2287 \text{ H}, \qquad C_{l2} = 1.231291\times10^{-2} \text{ F}$$

$$L_{l3} = 245.7739 \text{ H}, \qquad C_{l4} = 8.465688\times10^{-3} \text{ F}$$

$$L_{l5} = 120.2121 \text{ H}, \qquad C_{l6} = 1.825028\times10^{-3} \text{ F}$$

The desired maximally-flat band-elimination filter is given by

$$C_{e1} = \frac{1}{L_{l1}B} = 1.277376 \text{ nF}, \qquad L_{e1} = \frac{1}{\omega_0^2 C_{l1}} = 0.156571 \text{ mH}$$

$$L_{e2} = \frac{1}{C_{l2}B} = 24.610779 \text{ μH}, \qquad C_{e2} = \frac{1}{\omega_0^2 L_{l2}} = 8.126521 \text{ nF}$$

$$C_{e3} = \frac{1}{L_{l3}B} = 1.232964 \text{ nF}, \qquad L_{e3} = \frac{1}{\omega_0^2 C_{l3}} = 0.162211 \text{ mH}$$

$$L_{e4} = \frac{1}{C_{l4}B} = 35.795118 \text{ μH}, \qquad C_{e4} = \frac{1}{\omega_0^2 L_{l4}} = 5.587354 \text{ nF}$$

$$C_{e5} = \frac{1}{L_{l5}B} = 2.520797 \text{ nF}, \qquad L_{e5} = \frac{1}{\omega_0^2 C_{l5}} = 79.339986 \text{ μH}$$

$$L_{e6} = \frac{1}{C_{l6}B} = 0.166041 \text{ mH}, \qquad C_{e6} = \frac{1}{\omega_0^2 L_{l6}} = 1.204518 \text{ nF}$$

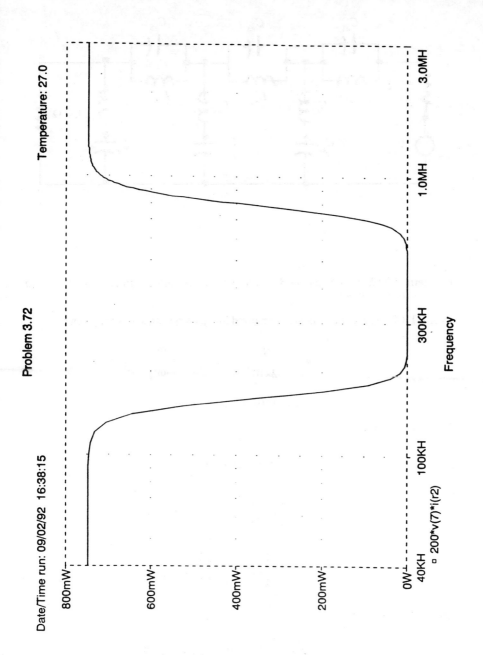

59

Problem 3.72

Date/Time run: 09/02/92 16:38:15

Temperature: 27.0

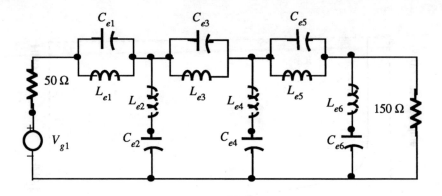

Problem 3.73 *Repeat Example 3.16 for a high-pass elliptic filter and plot its gain response as a function of ω.*

SOLUTION: The corresponding normalized low-pass specifications are given by

$$\omega'_c = 1, \qquad \omega'_s = \frac{10^5}{2\times10^4} = 5, \qquad k = \frac{\omega'_c}{\omega'_s} = \frac{1}{5} = 0.2$$

$$\varepsilon = 0.50844 \quad (1 \text{ dB ripple})$$

Choose $n = 4$. Then

$$k_1 = 2.6035\times10^5$$

$$A_{dB} = 10 \log\left(1 + \frac{\varepsilon^2}{k_1^2}\right) = 85.82 \text{ dB} > 60 \text{ dB}$$

$$H_4 = \left(1 + \varepsilon^2\right)\left[1 - \left(\frac{2-1}{2+1}\right)^2\right] = 1.12 > 1$$

Choose $H_4 = 1$. Then

$$\varepsilon = \sqrt{\frac{1}{8/9} - 1} = 0.35355 \quad (0.51 \text{ dB})$$

$$k_1 = 2.6\times10^{-5}, \qquad A_{dB} = 10\log\left(1 + \frac{\varepsilon^2}{k_1^2}\right) = 82.66\text{dB} > 60\text{ dB}$$

$$K_1 = 1.5708, \qquad K = 1.5869, \qquad \omega_1 = 0.386, \qquad \omega_2 = 0.9253$$

$$y_{p1} = -0.4233 \pm j0.4263, \qquad y_{p2} = -0.1705 \pm j1.0157$$

$$r(y) = y^4 + 1.1877y^3 + 1.7104y^2 + 1.0211y + 0.3828$$

$$\hat{r}(y) = y^4 + 1.005y^2 + 0.1275$$

$$\frac{Z_{11}(s)}{R_1} = \frac{2y^4 + 1.1877y^3 + 2.7154y^2 + 1.0211y + 0.5103}{1.1877y^3 + 0.7054y^2 + 1.0211y + 0.2553}$$

Problem 3.74 *Repeat Example 3.17 for a maximally-flat band-pass filter and plot its gain response as a function of* ω.

SOLUTION: Since

$$\omega_l = \frac{1}{B}\left(\omega_B - \frac{\omega_0^2}{\omega_B}\right)$$

we have

$$\omega_{rl} = \frac{1}{B}\left(4\times10^5 - \frac{4\times10^{10}}{4\times10^5}\right) = \frac{3\times10^5}{B}$$

$$\omega_{sl} = \frac{1}{B}\left(15.263\times10^5 - \frac{4\times10^{10}}{15.263\times10^5}\right) = \frac{15.000928\times10^5}{B}$$

where

$$\omega_0^2 = \omega_1\omega_2 = 4\times10^{10}$$

and B is unknown.

$$K_n = 1 - \left(\frac{R_2 - R_1}{R_2 + R_1}\right)^2 = 1 - \left(\frac{1}{3}\right)^2 = \frac{8}{9}$$

$$n \geq \frac{\log\left(10^{0.1 \times 60} - 1\right) - \log\left(10^{0.1} - 1\right)}{2 \log \dfrac{15.000928 \times 10^5}{3 \times 10^5}} = 4.7116 \quad \rightarrow \quad n = 5$$

Choose $n = 5$. Then we have $K_5 = 8/9$ and

$$\delta = \sqrt[5]{\frac{R_2 - R_1}{R_2 + R_1}} = \sqrt[5]{\frac{100}{300}} = 0.802742$$

The equivalent normalized low-pass response is given by

$$G\left(\omega^2\right) = \frac{K_5}{1 + \left(\dfrac{\omega_l}{\omega_c}\right)^2}$$

where $\omega_c = 1$, obtaining

$$\omega_{sl} = \left(10^{0.1 \times 60} - 1\right)^{\frac{1}{10}} = 3.981071307 = \frac{15.000928 \times 10^5}{B_s}$$

$$\rightarrow \quad B_s = 3.768063127 \times 10^5$$

$$\omega_{rl} = \left(10^{0.1 \times 1} - 1\right)^{\frac{1}{10}} = 0.873610 = \frac{3 \times 10^5}{B_r} \quad \rightarrow \quad B_r = 3.434028 \times 10^5$$

Choose

$$B = 3.6 \times 10^5 \text{ rad/s}$$

between B_r and B_s. The element values of the low-pass ladder network are found to be

$$L_{l1} = 313.3123 \text{ H}, \qquad C_{l2} = 9.237103 \times 10^{-3} \text{ F}, \qquad L_{l3} = 305.0964 \text{ H}$$

$$C_{l4} = 4.955221 \times 10^{-3} \text{ F}, \qquad L_{l5} = 68.5666 \text{ H}$$

The desired maximally-flat band-pass filter is obtained as follows:

$$L_{b1} = \frac{L_{l1}}{B} = 0.870312 \text{ mH}, \qquad C_{b1} = \frac{1}{L_{l1}\omega_0^2} = 28.725333 \text{ nF}$$

$$C_{b2} = \frac{C_{l2}}{B} = 25.658619 \text{ nF}, \qquad L_{b2} = \frac{1}{C_{l2}\omega_0^2} = 0.974331 \text{ mH}$$

$$L_{b3} = \frac{L_{l3}}{B} = 0.84749 \text{ mH}, \qquad C_{b3} = \frac{1}{L_{l3}\omega_0^2} = 29.498873 \text{ nF}$$

$$C_{b4} = \frac{C_{l4}}{B} = 13.764503 \text{ nF}, \qquad L_{b4} = \frac{1}{C_{l4}\omega_0^2} = 1.816266 \text{ mH}$$

$$L_{b5} = \frac{L_{l5}}{B} = 0.190463 \text{ mH}, \qquad C_{b5} = \frac{1}{L_{l5}\omega_0^2} = 0.131259 \text{ μF}$$

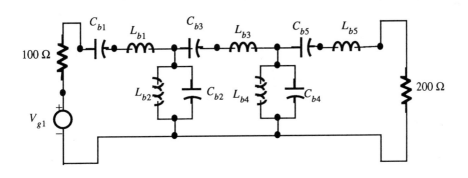

64

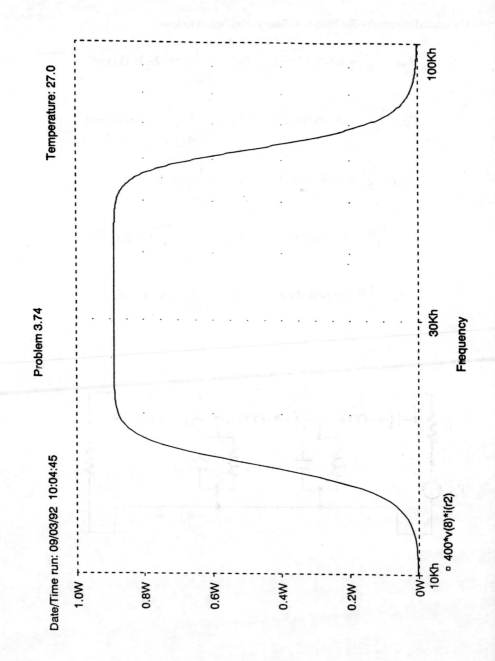

CHAPTER 4

THE PASSIVE LOAD

Problem 4.2 *Repeat the problem stated in Example 4.1 for n = 4.*

SOLUTION:

$$z_I(s) = \frac{1}{s+2}$$

$$r_I(s) = \frac{1}{2}\left(\frac{1}{s+2} + \frac{1}{-s+2}\right) = \frac{-2}{(s+2)(s-2)}$$

$$A(s) = \frac{s-2}{s+2}$$

$$F(s) = 2\, r_I(s)A(s) = \frac{-4}{(s+2)^2}$$

$$w(s) = \frac{r_I(s)}{z_I(s)} = \frac{-2}{s-2}$$

showing that it is a Class II zero of transmission of order 1.

$$A(s) = \frac{s-2}{s+2} = (s-2)(s+2)^{-1} = (s-2)\left(s^{-1} - 2s^{-2} + 4s^{-3} + \ldots\right)$$

$$= 1 - \frac{4}{s} + \frac{8}{s^2} + \ldots$$

$$F(s) = \frac{-4}{(s+2)^2} = -4\,(s+2)^{-2} = -4\left(s^{-2} - 4s^{-3} + \ldots\right)$$

$$= -\frac{4}{s^2} + \frac{16}{s^3} + \ldots$$

$$\rho(s)\rho(-s) = 1 - G\left(-s^2\right) = 1 - \frac{K_4}{1+s^8} = \frac{1 - K_4 + s^8}{1+s^8}$$

$$= \alpha^8 \frac{1 + y^8}{1 + s^8}$$

where

$$\alpha^8 = 1 - K_4, \qquad y = \frac{s}{\alpha}$$

$$\rho(s) = \alpha^4 \frac{y^4 + 2.61313y^3 + 3.41421y^2 + 2.61313y + 1}{s^4 + 2.61313s^3 + 3.41421s^2 + 2.61313s + 1}$$

$$= \frac{s^4 + 2.61313\alpha s^3 + 3.41421\alpha^2 s^2 + 2.61313\alpha^3 s + \alpha^4}{s^4 + 2.61313s^3 + 3.41421s^2 + 2.61313s + 1}$$

$$= \rho_0 + \frac{\rho_1}{s} + \frac{\rho_2}{s^2} + \ldots$$

$$\rho_0 = \rho(\infty) = 1$$

$$\rho_1 = s\left[\rho(s) - \rho(\infty)\right]\big|_{s=\infty} = 2.61313(\alpha - 1)$$

$$A_0 = 1 = \rho_0$$

$$\frac{A_1 - \rho_1}{F_2} \geq 0 \quad \rightarrow \quad \frac{-4 - 2.61313(\alpha - 1)}{-4} \geq 0$$

$$4 + 2.61313(\alpha - 1) \geq 0$$

$$\alpha \geq -0.5307$$

Choose α as small as possible (α is bounded between 0 and 1) by letting $\alpha = 0$, obtaining

$$K_4 = 1$$

Alternatively, since

$$\frac{2 \sin \frac{\pi}{2n}}{RC\omega_c} = 4 \sin \frac{\pi}{8} = 1.531 > 1$$

$K_4 = 1$ is possible.

$$Z_{22}(s) = \frac{F(s)}{A(s) - \rho(s)} - z_l(s)$$

$$= \frac{\dfrac{-4}{(s+2)^2}}{\dfrac{s-2}{s+2} - \dfrac{s^4}{s^4 + 2.61313s^3 + 3.41421s^2 + 2.61313s + 1}} - \frac{1}{s+2}$$

$$= \frac{2.6131s^3 + 3.4143s^2 + 2.6130s + 1}{1.38687s^4 + 1.81205s^3 + 4.21529s^2 + 4.22626s + 2}$$

$$= \cfrac{1}{0.5307s + \cfrac{1}{0.92387s + \cfrac{1}{3.69576s + \cfrac{1}{0.38266s + \dfrac{1}{2}}}}}$$

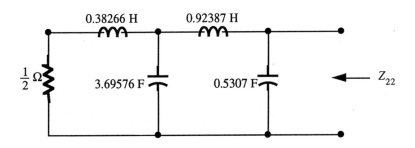

Magnitude-scaling it back by a factor of 100 and frequency-scaling it back by 10^8, we obtain the desired matching network below.

68

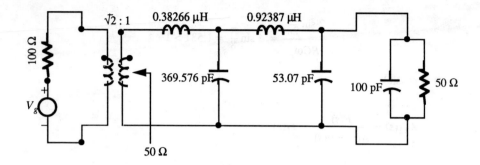

Problem 4.3 *Repeat the problem given in Example 4.2 for a 2-dB passband tolerance.*

SOLUTION:

$$10 \log (1 + \varepsilon^2) = 2$$

$$\varepsilon = \sqrt{10^{0.2} - 1} = 0.76478$$

$$\frac{2 \sin \dfrac{\pi}{2n}}{RC\omega_c} = \sin \frac{\pi}{8} = 0.3827 > \sinh a = \sinh \left(\frac{1}{4} \sinh^{-1} \frac{1}{0.76478} \right) = 0.27408$$

showing that $K_4 = 1$ is attainable.

$$\rho(s)\rho(-s) = \frac{\varepsilon^2 C_n^2(-js)}{1 + \varepsilon^2 C_n^2(-js)} = \frac{\varepsilon^2 \left(8s^4 + 8s^2 + 1 \right)^2}{1 + \varepsilon^2 \left(8s^4 + 8s^2 + 1 \right)^2}$$

$$= \frac{\varepsilon^2 8^2 \left(s^4 + s^2 + 0.125 \right) \left(s^4 + s^2 + 0.125 \right)}{1 + \varepsilon^2 \left(8s^4 + 8s^2 + 1 \right)^2}$$

$$\rho(s) = \frac{s^4 + s^2 + 0.125}{s^4 + 0.71622s^3 + 1.25648s^2 + 0.51680s + 0.20577}$$

$$z_l(s) = \frac{1}{1 + 2s}$$

$$z_l(-s) = \frac{1}{1 - 2s}$$

$$A(s) = \frac{s - \frac{1}{2}}{s + \frac{1}{2}} = \frac{2s - 1}{2s + 1}$$

$$r_l(s) = \frac{1}{2}\left[z_l(s) + z_l(-s)\right] = \frac{1}{2}\left(\frac{1}{1 + 2s} + \frac{1}{1 - 2s}\right) = \frac{1}{1 - 4s^2}$$

$$F(s) = 2\,r_l(s)A(s) = 2 \times \frac{1}{(1 + 2s)(1 - 2s)} \times \frac{2s - 1}{2s + 1} = \frac{-2}{(2s + 1)^2}$$

$$Z_{22}(s) = \frac{F(s)}{A(s) - \rho(s)} - z_l(s)$$

$$= \frac{\dfrac{-2}{(2s + 1)^2}}{\dfrac{2s - 1}{2s + 1} - \dfrac{s^4 + s^2 + 0.125}{s^4 + 0.71622s^3 + 1.25648s^2 + 0.51680s + 0.20577}} - \frac{1}{2s + 1}$$

$$= \frac{1.43244s^4 + 1.22918s^3 + 1.29008s^2 + 0.67834s + 0.08077}{(2s + 1)(0.56756s^4 + 0.20326s^3 + 1.22288s^2 + 0.35526s + 0.33077)}$$

$$= \frac{0.71722s^3 + 0.25648s^2 + 0.5168s + 0.08077}{0.56756s^4 + 0.20326s^3 + 1.22288s^2 + 0.35526s + 0.33077}$$

$$= \cfrac{1}{0.79244s + \cfrac{1}{0.88058s + \cfrac{1}{3.6064s + \cfrac{1}{0.68183s + 0.24419}}}}$$

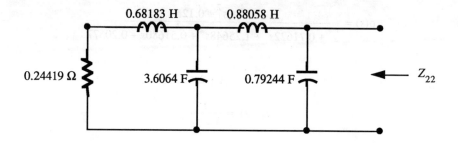

Magnitude-scaling it by a factor of $b = 10^2$ and frequency-scaling it by a factor of $a = 10^8$, we obtain the desired matching network below.

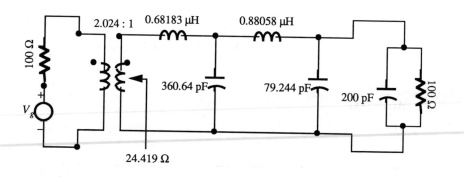

Problem 4.4 *For n =1, the minimum-phase solution $\hat{\rho}(s)$ of (4.17a) becomes*

$$\hat{\rho}(s) = \frac{s + (1 - K_1)^{1/2}\omega_c}{s + \omega_c}$$

Using this, derive the equalizer back-end driving-point impedance $Z_{22}(s)$ and realize this impedance as a lossless two-port network terminated in a 1-Ω resistor.

SOLUTION: **Case 1.**

$$\frac{2}{RC\omega_c} \geq 1$$

Then we have $K_1 = 1$ and

$$\rho(s) = \hat{\rho}(s) = \frac{s}{s + \omega_c}$$

$$Z_{22}(s) = \frac{F(s)}{A(s) - \rho(s)} - z_l(s) = \frac{\dfrac{-2\tau}{C}}{(s+\tau)^2} - \dfrac{\dfrac{1}{C}}{s+\tau}$$

$$= \frac{-\dfrac{2\tau}{C}(s + \omega_c)}{(s + \tau)\left[(\omega_c - 2\tau)s - \omega_c\tau\right]} - \frac{\dfrac{1}{C}}{s + \tau}$$

$$= \frac{\dfrac{\omega_c}{C}}{(2\tau - \omega_c)s + \omega_c\tau}$$

where $\tau = 1/RC$, since

$$2\tau - \omega_c \geq 0$$

or

$$\frac{2}{RC\omega_c} \geq 1$$

Thus, $Z_{22}(s)$ is realizable.

Case 2.

$$\frac{2}{RC\omega_c} < 1$$

Then we have

$$K_1 = 1 - \left(1 - \frac{2}{RC\omega_c}\right)^2$$

$$\rho(s) = \hat{\rho}(s) = \frac{s + \left(1 - \dfrac{2}{RC\omega_c}\right)\omega_c}{s + \omega_c} = \frac{s + \left(\omega_c - \dfrac{2}{RC}\right)}{s + \omega_c}$$

$$Z_{22}(s) = \frac{F(s)}{A(s) - \hat{\rho}(s)} - z_l(s) = \frac{\dfrac{-2\tau}{C}}{\dfrac{(s+\tau)^2}{\dfrac{s-\tau}{s+\tau} - \dfrac{s + (\omega_c - 2\tau)}{s + \omega_c}}} - \frac{\dfrac{1}{C}}{s+\tau}$$

$$= \frac{\dfrac{-2\tau}{C}(s + \omega_c)}{(s+\tau)\left[(s-\tau)(s+\omega_c) - (s+\tau)(s+\omega_c - 2\tau)\right]} - \frac{\dfrac{1}{C}}{s+\tau}$$

$$= \frac{\dfrac{-2\tau}{C}(s + \omega_c)}{2\tau(s+\tau)(\tau - \omega_c)} - \frac{\dfrac{1}{C}}{s+\tau} = \frac{1}{C(\omega_c - \tau)} = \frac{R}{RC\omega_c - 1} > 0$$

Problem 4.10 *It is required to equalize the parallel combination of a 60-Ω resistor and a 150-pF capacitor to a resistive generator of internal resistance 100 Ω, and to achieve the fourth-order low-pass Butterworth transducer power gain with maximal attainable dc gain. The 3-dB bandwidth is 10^8 rad/s Realize the desired lossless equalizer.*

SOLUTION:

Step 1.

$$G(\omega^2) = \frac{K_4}{1 + \omega^8}, \qquad 0 \le K_4 \le 1$$

Step 2. Frequency-scaling the elements by a factor of 10^8 and magnitude-scaling them by a factor 100 yield

$$z_l(s) = \frac{1}{1.5s + \dfrac{1}{0.6}} = \frac{6}{9s + 10}$$

$$r_l(s) = \frac{1}{2}\left[z_l(s) + z_l(-s)\right] = \frac{1}{2}\left(\frac{6}{9s+10} + \frac{6}{-9s+10}\right) = \frac{60}{(9s+10)(10-9s)}$$

$$A(s) = \frac{s - \frac{10}{9}}{s + \frac{10}{9}} = \frac{9s - 10}{9s + 10}$$

$$F(s) = 2\, r_1(s) A(s) = 2 \times \frac{60}{(9s + 10)(10 - 9s)} \times \frac{9s - 10}{9s + 10} = \frac{-120}{(9s + 10)^2}$$

Step 3.

$$\frac{r_1(s)}{z_1(s)} = \frac{60(9s + 10)}{6(9s + 10)(10 - 9s)} = \frac{10}{10 - 9s}$$

Thus, $s = \infty$ is a Class II zero of transmission of order $k = 1$.

Step 4.

$$\rho(s)\rho(-s) = 1 - G\!\left(-s^2\right) = 1 - \frac{K_4}{1 + s^8} = \alpha^8 \frac{1 + y^8}{1 + s^8}$$

where

$$\alpha^8 = 1 - K_4, \qquad y = \frac{s}{\alpha}$$

$$\rho(s) = \alpha^4 \frac{y^4 + 2.6131 y^3 + 3.4142 y^2 + 2.6131 y + 1}{s^4 + 2.6131 s^3 + 3.4142 s^2 + 2.6131 s + 1}$$

Step 5.

$$A(s) = 1 - \frac{\frac{20}{9}}{s} + \dots$$

$$F(s) = 0 + 0 - \frac{\frac{120}{81}}{s^2} + \dots$$

$$\rho(s) = 1 + \frac{\rho_1}{s} + \dots$$

where

$$\rho_1 = 2.6131(\alpha - 1)$$

Step 6. For a Class II zero of transmission of order $k = 1$, the coefficient constraints become

$$A_0 = \rho_0, \qquad \frac{A_1 - \rho_1}{F_2} \geq 0$$

From Step 5,

$$A_0 = 1 \text{ and } \rho_0 = 1 \rightarrow A_0 = \rho_0$$

$$\frac{A_1 - \rho_1}{F_2} = \frac{-\frac{20}{9} - 2.6131(\alpha - 1)}{-\frac{120}{81}} \geq 0$$

giving

$$2.6131(\alpha - 1) \geq -\frac{20}{9} \quad \rightarrow \quad 2.6131\alpha \geq 2.6131 - \frac{20}{9}$$

or

$$\alpha \geq 0.1495839$$

From

$$\alpha^8 = 1 - K_4, \qquad 0 \leq K_4 \leq 1$$

we see that K_4 should be close to unity. Thus, choose

$$\alpha = 0.149584, \quad \left(K_4 = 1 - \alpha^8 = 0.9999997\right)$$

obtaining

$$\rho(s) = \frac{s^4 + 0.3909s^3 + 0.0764s^2 + 0.0087s + 0.0005}{s^4 + 2.6131s^3 + 3.4142s^2 + 2.6131s + 1}$$

Step 7.

$$Z_{22}(s) = \frac{F(s)}{A(s) - \rho(s)} - z_1(s) = \frac{\dfrac{-120}{(9s + 10)^2}}{\dfrac{9s - 10}{9s + 10} - \rho(s)} - \frac{6}{9s + 10}$$

$$= \frac{13.3333s^3 + 20.0268s^2 + 15.6263s + 5.9970}{11.4667s^2 + 17.2233s + 10.005}$$

Step 8.

$$Z_{22}(s) = 1.163s + \cfrac{1}{2.872s + \cfrac{1}{0.399s + 0.5994}}$$

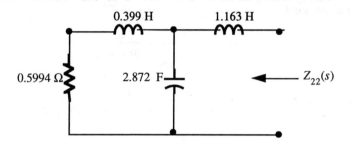

Denormalizing the elements using frequency-scaling by a factor 10^8 and magnitude-scaling by a factor of 100 yields the following desired matching network:

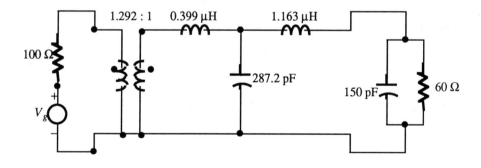

Problem 4.11 *Consider the same problem as in Problem 4.10 except that now we wish to achieve the third-order Chebyshev transducer power gain having a maximum attainable K_3. The passband tolerance is 1.5 dB and the cutoff frequency is $50/\pi$ MHz. Design a lossless equalizer with the desired properties.*

SOLUTION:

Step 1.

$$G(\omega^2) = \frac{K_3}{1 + \varepsilon^2 C_3^2(\omega)}, \qquad 0 \le K_3 \le 1$$

where $C_3(\omega) = 4\omega^3 - 3\omega$.

Step 2. Frequency-scaling the elements by a factor of 10^8 and magnitude-scaling them by a factor of 100 yield

$$z_I(s) = \frac{1}{1.5s + \dfrac{1}{0.6}} = \frac{6}{9s + 10}$$

$$r_I(s) = \frac{1}{2}\left[z_I(s) + z_I(-s)\right] = \frac{1}{2}\left(\frac{6}{9s + 10} + \frac{6}{-9s + 10}\right) = \frac{60}{(9s + 10)(10 - 9s)}$$

$$A(s) = \frac{s - \dfrac{10}{9}}{s + \dfrac{10}{9}} = \frac{9s - 10}{9s + 10}$$

$$F(s) = 2\,r_I(s)A(s) = \frac{-120}{(9s + 10)^2}$$

Step 3.

$$\frac{r_I(s)}{z_I(s)} = \frac{10}{10 - 9s}$$

Thus, $s = \infty$ is a Class II zero of transmission of order $k = 1$.

Step 4.

$$\rho(s)\rho(-s) = 1 - G(-s^2) = 1 - \frac{K_3}{1 + \varepsilon^2 C_3^2(-js)} = \frac{1 - K_3 + \varepsilon^2 C_3^2(-js)}{1 + \varepsilon^2 C_3^2(-js)}$$

$$\rho(s) = \frac{s^3 + a_1 s^2 + a_2 s + a_3}{s^3 + 0.98834s^2 + 1.23841s + 0.49131}$$

Step 5.

$$A(s) = 1 - \frac{\frac{20}{9}}{s} + \dots$$

$$F(s) = 0 + 0 - \frac{\frac{120}{81}}{s^2} + \dots$$

$$\rho(s) = 1 + \frac{a_1 - 0.98834}{s} + \dots$$

Step 6. For a Class II zero of transmission of order $k = 1$, the coefficient constraints become

$$A_0 = \rho_0, \qquad \frac{A_1 - \rho_1}{F_2} \geq 0$$

From Step 5, we obtain

$$A_0 = 1 \text{ and } \rho_0 = 1 \quad \rightarrow \quad A_0 = \rho_0$$

$$\frac{A_1 - \rho_1}{F_2} = \frac{-\frac{20}{9} - (a_1 - 0.98834)}{-\frac{120}{81}} \geq 0$$

or

$$a_1 \geq 0.98834 - \frac{9}{20} \quad \text{or} \quad a_1 \geq -1.23388$$

Choose $K_3 = 1$.

$$\varepsilon\, C_3(-js) = \varepsilon\left[4(-js)^3 - 3(-js) \right] = 4\,\varepsilon j\left(s^3 + 0.75s \right)$$

where

$$C_3(\omega) = 4\omega^3 - 3\omega$$

obtaining

$$\rho(s)\rho(-s) = \frac{1 - K_3 + \varepsilon^2 C_3^2(-js)}{1 + \varepsilon^2 C_3^2(-js)} = \frac{\varepsilon^2 C_3^2(-js)}{1 + \varepsilon^2 C_3^2(-js)}$$

$$\Rightarrow \quad \rho(s) = \frac{s^3 + 0.75s}{s^3 + 0.98834s^2 + 1.23841s + 0.49131}$$

showing that $a_1 = 0$ and $a_3 = 0$.

Step 7.

$$Z_{22}(s) = \frac{F(s)}{A(s) - \rho(s)} - z_l(s) = \frac{\dfrac{-120}{(9s + 10)^2}}{\dfrac{9s - 10}{9s + 10} - \rho(s)} - \frac{6}{9s + 10}$$

$$= \frac{5.93004s^2 + 2.93046s + 2.94786}{11.10494s^3 + 5.48771s^2 + 15.46231s + 4.9131}$$

$$= \cfrac{1}{1.873s + \cfrac{1}{0.596s + \cfrac{1}{3.373s + \dfrac{4.913}{2.948}}}}$$

Step 8.

$$Z_{22}(s) = \cfrac{1}{1.873s + \cfrac{1}{0.5963s + \cfrac{1}{3.373s + \dfrac{1}{0.60}}}}$$

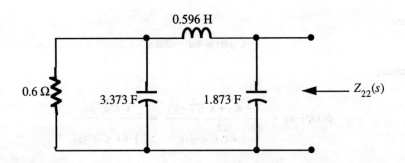

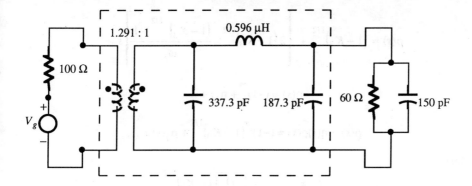

Problem 4.17 *Design a lossless matching network to equalize a load composed of a series connection of an R-ohm resistor and a C-farad capacitor to a resistive generator and to achieve the third-order Butterworth transducer power gain of low-pass type. The radian cutoff frequency is ω_c. Derive the equalizer back-end bounded-real reflection coefficient and the corresponding impedance.*

SOLUTION:

$$z_l(s) = R + \frac{1}{sC} = \frac{RCs + 1}{sC}, \qquad z_l(-s) = R - \frac{1}{sC} = \frac{RCs - 1}{sC}$$

$$r_l(s) = \frac{1}{2}\left[z_l(s) + z_l(-s)\right] = R$$

$$w(s) = \frac{r_l(s)}{z_l(s)} = \frac{RCs}{RCs + 1} = \frac{s}{s + \dfrac{1}{RC}}$$

Thus, $s_0 = 0 + j0$ and $z_l(0) = \infty$ imply that $s_0 = 0$ is a Class IV zero of transmission of order $k = 1$. The coefficient constraints become

$$A_0 = \rho_0, \qquad \frac{F_0}{A_1 - \rho_1} \geq \frac{1}{C}$$

$$A(s) = 1$$

$$F(s) - 2r_l(s)A(s) = 2R$$

$$\hat{\rho}(s) = \left(1 - K_3\right)^{1/2} + \left[2\left(1 - K_3\right)^{1/2} \times \frac{\left(1 - K_3\right)^{-1/2} - 1}{\omega_c}\right] s + \dots$$

$$\eta(s) = (-1)^m + \eta_1 s + \dots$$

$$\rho(s) = \eta(s)\hat{\rho}(s) = (-1)^m\left(1 - K_3\right)^{1/2} + \rho_1(s) + \dots$$

Since

$$A_0 = 1 \neq \rho_0 = (-1)^m\left(1 - K_3\right)^{1/2}$$

unless $K_3 = 0$, there is no solution.

Problem 4.18 *Repeat the problem given in Example 4.11 for the steepness* $1/k = 1.3$, *everything else being the same.*

SOLUTION:

$$k = \frac{1}{1.3} = 0.769231, \qquad K = 1.940717$$

$$\sigma_0 = \frac{1}{R_2 C}\left(1 + \frac{R_2}{R_1}\right)^{1/2} = 10^8, \qquad y_0 = \frac{\sigma_0}{\omega_c} = 1$$

$$A(s) = \frac{R_2 Cs - 1}{R_2 Cs + 1} = \frac{2\times10^{-8}s - 1}{2\times10^{-8}s + 1}, \qquad A(\sigma_0) = \frac{1}{3}$$

For $n = 3$, $\varepsilon = 0.34931$, and $k = 0.769231$, we have

$$r(y) = y^3 + 1.2248820y^2 + 1.4631437y + 0.9734109$$

$$r(y_0) = 4.6614366$$

$$\hat{r}_m(s) = s\left(s^2 + 0.826182\right)$$

$$\hat{r}_m(y_0) = 1.826182$$

where $y = s/10^8$, giving

$$A(\sigma_0)r(y_0) = 1.5538122 < 1.826182 = \lambda \hat{r}_m(y_0)$$

Thus, Case 1 applies, $H_3 = 1$ and we have to insert an open RHS zero in $\rho(s)$. The desired zero is located at

$$\sigma_1 = \sigma_0 \frac{\lambda \hat{r}_m(y_0) - A(\sigma_0)r(y_0)}{\lambda \hat{r}_m(y_0) + A(\sigma_0)r(y_0)} = 10^8 \frac{1.826182 - 1.5538122}{1.826182 + 1.5538122} = 0.08058 \times 10^8$$

$$\eta(s) = \frac{s - \sigma_1}{s + \sigma_1}$$

$$\rho(y) = \eta(y)\hat{\rho}(y) = \frac{y(y - 0.08058)(y^2 + 0.826182)}{(y + 0.08058)\left(y^3 + 1.224882y^2 + 1.4631437y + 0.97341\right)}$$

$$= \frac{y^4 - 0.08058y^3 + 0.82618y^2 - 0.06657y}{y^4 + 1.30546y^3 + 1.5618y^2 + 1.0913y + 0.07844}$$

Using (4.220) yields

$$Z_{22}(y) = \frac{400y^5 + 244.976y^4 - 76.82y^3 - 89.3y^2 - 447.46y - 31.376}{0.772y^4 + 0.24636y^3 - 0.07224y^2 - 0.8678y - 0.07844}$$

$$= \frac{400y^4 + 644.98y^3 + 568.16y^2 + 478.86y + 31.4}{0.772y^3 + 1.0184y^2 + 0.946y + 0.078}$$

$$Z_{22}(s) = 5.18 \times 10^{-6} s + \cfrac{1}{9.9 \times 10^{-11} s + \cfrac{1}{1.23 \times 10^{-6} s + \cfrac{1}{2.02 \times 10^{-10} s + \cfrac{1}{400}}}}$$

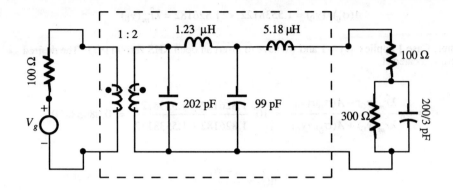

Problem 4.19 *Repeat the problem given in Example 4.11 for n = 4.*

SOLUTION:

$$k = \frac{1}{1.4} = 0.7142857$$

$$\sigma_0 = 10^8, \quad y_0 = 1$$

$$A(s) = \frac{2 \times 10^{-8} s - 1}{2 \times 10^{-8} s + 1}, \qquad A(s_0) = \frac{1}{3}$$

For $n = 4$, $k = 0.7142857$ and $\varepsilon = 0.34931$, we have

$$r(y) = y^4 + 1.1705639y^3 + 1.7798206y^2 + 1.1070699y + 0.5287467$$

$$r(y_0) = 5.5862011$$

$$\hat{r}_m(s) = \left(s^2 + 0.1950658\right)\left(s^2 + 0.0.8938979\right)$$

$$\hat{r}_m(y_0) = 2.26333$$

where $y = s/10^8$, giving

$$A(\sigma_0)r(y_0) = 1.862067 \ < \ 2.26333 = \lambda \hat{r}_m(y_0)$$

$$\sigma_1 = 10^8 \frac{2.26333 - 1.862067}{2.26333 + 1.862067} = 0.09727 \times 10^8$$

$$\rho(y) = \eta(y)\hat{\rho}(y) = \frac{(y - 0.09727)\left(y^2 + 0.1950658\right)\left(y^2 + 0.8938979\right)}{(y + 0.0972)\left(y^4 + 1.1705639y^3 + 1.779821y^2 + 1.10707y + 0.5287\right)}$$

$$= \frac{y^5 - 0.09727y^4 + 1.089y^3 - 0.1059y^2 + 0.1744y - 0.01696}{y^5 + 1.2678y^4 + 1.8937y^3 + 1.2802y^2 + 0.6364y + 0.05143}$$

Problem 4.21 *Repeat Problem 4.17 for the load composed of a series connection of an R-ohm resistor and an L-henry inductor.*

SOLUTION:

$$z_1(-s) = R - sL$$

$$r_1(s) = \frac{1}{2}\left[z_1(s) + z_1(-s)\right] = R$$

$$w(s) = \frac{r_1(s)}{z_1(s)} = \frac{R}{R + sL} \quad \rightarrow \quad s_0 = \infty, \quad k = 1$$

Since $z_1(s_0) = \infty$, s_0 is a Class IV zero of transmission of order 1. The coefficient constraints become

$$A_0 = \rho_0, \qquad \frac{F_0}{A_1 - \rho_1} \geq a_{-1}$$

$$A(s) = -1, \qquad F(s) = 2\,r_1(s)A(s) = -2R$$

Choose $\rho_0 = -1$. Then we have

$$\rho_1 = + \frac{\left[1 - \left(1 - K_3\right)^{1/6}\right]\omega_c}{\sin\frac{\pi}{6}} = + 2\omega_c\left[1 - \left(1 - K_3\right)^{1/6}\right]$$

$$\frac{-2R}{0 - 2\omega_c\left|1 - \left(1 - K_3\right)^{1/6}\right|} \geq a_{-1} = L \quad \text{or} \quad \left(1 - K_3\right)^{1/6} \geq 1 - \frac{R}{\omega_c L}$$

Case 1. $R \geq \omega_c L$ and $K_3 = 1$.

$$\rho(s) = \frac{y^3}{y^3 + 2y^3 + 2y + 1} = \frac{s^3}{s^3 + 2\omega_c s^2 + 2\omega_c^2 s + \omega_c^3}$$

where $y = s/\omega_c$ and $\omega_c = 10^8$.

$$Z_{22}(s) = \frac{F(s)}{A(s) - \rho(s)} - z_l(s)$$

$$= \frac{(R - \omega_c L)s^3 + \omega_c(R - \omega_c L)s^2 + \omega_c^2(R - 0.5\omega_c L)s + 0.5\omega_c^3}{\omega_c s^2 + \omega_c^2 s + 0.5\omega_c^3}$$

Case 2.

$$R < \omega_c L, \qquad K_3 \leq 1 - \left(1 - \frac{R}{\omega_c L}\right)^6$$

Choose

$$K_3 = 1 - \left(1 - \frac{R}{\omega_c L}\right)^6$$

$$\rho(s) = -\left(1 - K_3\right)^{1/2} \frac{x^3 + 2x^2 + 2x + 1}{y^3 + 2y^2 + 2y + 1} = -\frac{y^3 + 2\delta y^2 + 2\delta^2 y + \delta^3}{y^3 + 2y^2 + 2y + 1}$$

where

$$x = \frac{y}{\delta}, \qquad \delta = \left(1 - K_3\right)^{1/6}, \qquad \alpha = \frac{\omega_c L}{R}$$

$$Z_{22}(s) = \frac{-2R}{-1 + \dfrac{y^3 + 2\delta y^2 + 2\delta^2 y + \delta^3}{y^3 + 2y^2 + 2y + 1}} - (R + \omega_c L y)$$

$$= \frac{R\left\{\alpha(1-\delta)y^3 + \left[\alpha\left(1-\delta^2\right)+(1-\delta)\right]y^2 + \left[\left(1-\delta^2\right)+0.5\left(1-\delta^3\right)\alpha\right]y + 0.5\left(1-\delta^3\right)\right\}}{(1-\delta)y^2 + \left(1-\delta^2\right)y + 0.5\left(1-\delta^3\right)}$$

$$Z_{22}(y) = \frac{400y^6 + 234.106y^5 + 50.512y^4 - 87.02y^3 - 392.28y^2 - 177.906y - 27.356}{0.73014y^5 + 0.43887y^4 - 0.2105y^3 - 0.2503y^2 - 0.674y - 0.03447}$$

$$= \frac{400y^5 + 634y^4 + 685y^3 + 598y^2 + 206y + 28}{0.73y^4 + 1.169y^3 + 0.9585y^2 + 0.7082y + 0.0342}$$

$$Z_{22}(s) = 5.48\times10^{-6}s + \cfrac{1}{4.568\times10^{-11}s + \cfrac{1}{15.5\times10^{-6}s + \cfrac{1}{7.676\times10^{-12}s + \cfrac{1}{39.24\times10^{-6}s + 819}}}}$$

which can be identified as an LC ladder network terminating in a 100-Ω resistor using the ideal transformer with turns ratio

$$\frac{1}{n} = \sqrt{\frac{100}{819}} = \frac{1}{2.86}$$

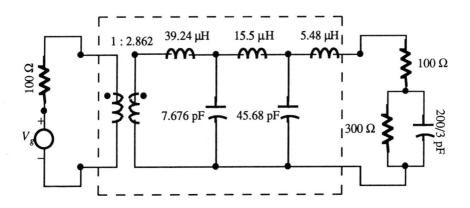

Problem 4.23 *It is desired to design a lossless matching network to equalize the load*

$$z_l(s) = s + \frac{1}{s+1}$$

to a resistive generator and to achieve the third-order Butterworth transducer power gain having a maximum attainable dc gain K_3. The normalized radian cutoff frequency is $\omega_c = 1$. Show that

(i) $K_3 = 1$ *can always be achieved without the insertion of the open RHS zeros in $\rho(s)$,*
(ii) *the corresponding equalizer back-end impedance $Z_{22}(s) = z_l(s)$,*
(iii) *the scattering matrix of the resulting equalizer realized as a lossless ladder terminated in a 1-Ω resistor is given by*

$$\mathbf{S}(s) = \frac{1}{(s+1)(s^2+s+1)} \begin{bmatrix} -s^3 & 1 \\ 1 & -s^3 \end{bmatrix}$$

normalizing to the reference impedances 1 and $z_l(s)$.

SOLUTION:

(i) We show that $K_3 = 1$ can always be achieved without the insertion of the open RHS zeros in $\rho(s)$.

Step 1.

$$0 \le G\left(\omega^2\right) \le 1 \text{ for all } \omega$$

$$0 \le K_3 \le 1$$

Step 2.

$$z_l(s) = s + \frac{1}{s+1} = \frac{s^2+s+1}{s+1}$$

$$r_l(s) = \frac{1}{2}\left[z_l(s) + z_l(-s)\right] = \frac{1}{2}\left(s + \frac{1}{s+1} - s + \frac{1}{-s+1}\right) = \frac{1}{1-s^2}$$

$$A(s) = \frac{s-1}{s+1}$$

$$F(s) = 2\,r_l(s)A(s) = 2 \times \frac{1}{1-s^2} \times \frac{s-1}{s+1} = \frac{-2}{(s+1)^2}$$

Step 3.

$$\frac{r_l(s)}{z_l(s)} = \frac{1}{1-s^2} \times \frac{s+1}{s^2+s+1} = \frac{1}{(1-s)(s^2+s+1)}$$

$s_0 = \infty$ is a Class IV zero of transmission of order $k = 3$.

Step 4.

$$\rho(s)\rho(-s) = 1 - G\left(-s^2\right) = 1 - \frac{K_3}{1-s^6} = \alpha^6 \frac{1-y^6}{1-s^6}$$

where

$$\alpha^6 = 1 - K_3, \qquad y = \frac{s}{\alpha}$$

$$\rho(s)\rho(-s) = \alpha^3 \times \frac{y^3 + 2y^2 + 2y + 1}{s^3 + 2s^2 + 2s + 1} = \frac{s^3 + 2\alpha s^2 + 2\alpha^2 s + \alpha^3}{s^3 + 2s^3 + 2s + 1}$$

Step 5.

$$A(s) = 1 - \frac{2}{s} + \frac{2}{s^2} - \frac{2}{s^3} + \dots$$

$$F(s) = 0 + \frac{0}{s} - \frac{2}{s^2} + \frac{4}{s^3} + \dots$$

$$\rho(s) = 1 + \frac{\rho_1}{s} + \frac{\rho_2}{s^2} + \frac{\rho_3}{s^3} + \dots$$

where

$$\rho_1 = 2(\alpha - 1), \qquad \rho_2 = 2(\alpha - 1)^2, \qquad \rho_3 = (\alpha - 1)\left(\alpha^2 - 3\alpha + 1\right)$$

Step 6. For a Class IV zero of transmission of order $k = 3$, the coefficient constraints become

$$A_0 = \rho_0, \quad A_1 = \rho_1, \quad A_2 = \rho_2, \quad \frac{F_2}{A_3 - \rho_3} \geq a_{-1} = 1$$

where

$$a_{-1} = \lim_{s \to -1} (s+1)z_l(s) = \lim_{s \to -1} (s^2 + s + 1) = 1$$

From Step 5 we have

$$A_0 = 1, \quad \rho_0 = 1 \quad \rightarrow \quad A_0 = \rho_0 \quad \text{O.K.}$$

Also

$$A_1 = -2, \quad \rho_1 = 2(\alpha - 1)$$

imply that if $\alpha = 0$ then

$$A_1 = -2, \quad \rho_1 = -2 \quad \text{O.K.}$$

and

$$A_2 = 2, \quad \rho_2 = 2(\alpha - 1)^2$$

or

$$A_2 = 2, \quad \rho_2 = 2 \quad \text{O.K.}$$

Finally,

$$A_3 = -2, \quad \rho_3 = (\alpha - 1)(\alpha^2 - 3\alpha + 1), \quad F_2 = -2$$

imply that

$$A_3 = -2, \quad \rho_3 = -1, \quad F_2 = -2$$

or

$$\frac{F_2}{A_3 - \rho_3} = \frac{-2}{-2 - (-1)} = 2 > 1 \quad \text{O.K.}$$

From the above, if we choose $\alpha = 0$, we see that the coefficient constraints can all be satisfied. Therefore, we need not consider the general bounded-real reflection coefficient

$$\rho(s) = \pm \eta(s)\hat{\rho}(s)$$

As a result, there is no need to insert the open RHS zeros in $\rho(s)$, and we set $\alpha = 0$. From Step 4 we have

$$\alpha^6 = 1 - K_3 \quad \rightarrow \quad 0 = 1 - K_3 \text{ or } K_3 = 1$$

Thus, $K_3 = 1$ can always be achieved without the insertion of the open RHS zeros in $\rho(s)$. From Step 4

$$\rho(s) = \left. \frac{s^3 + 2\alpha s^2 + 2\alpha^2 s + \alpha^3}{s^3 + 2s^2 + 2s + 1} \right|_{\alpha = 0} = \frac{s^3}{s^3 + 2s^2 + 2s + 1}$$

(ii) We show that $Z_{22}(s) = z_1(s)$.

Step 7. The equalizer back-end impedance $Z_{22}(s)$ is found to be

$$Z_{22}(s) = \frac{F(s)}{A(s) - \rho(s)} - z_I(s) = \frac{\dfrac{-2}{(s+1)^2}}{\dfrac{s-1}{s+1} - \dfrac{s^3}{s^3 + 2s^2 + 2s + 1}} - \frac{s^2 + s + 1}{s+1}$$

$$= \frac{-2(s^3 + 2s^2 + 2s + 1)}{(s+1)\left[(s-1)(s^3 + 2s^2 + 2s + 1) - s^3(s+1)\right]} - \frac{s^2 + s + 1}{s+1}$$

$$= \frac{-2(s^2 + s + 1)}{-s - 1} - \frac{s^2 + s + 1}{s+1} = \frac{s^2 + s + 1}{s+1}$$

Step 8.

$$Z_{22}(s) = s + \frac{1}{s+1}, \qquad z_I(s) = s + \frac{1}{s+1}$$

or

$$Z_{22}(s) = z_I(s)$$

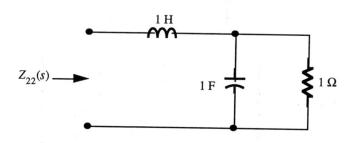

(iii) We show that

$$S(s) = \frac{1}{(s+1)(s^2 + s + 1)} \begin{bmatrix} -s^3 & 1 \\ 1 & -s^3 \end{bmatrix}$$

From Step 8, we have

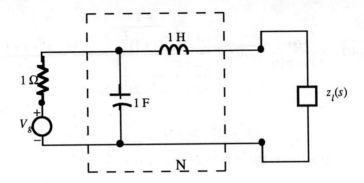

$$\mathbf{Z}(s) = \begin{bmatrix} \dfrac{1}{s} & \dfrac{1}{s} \\[3mm] \dfrac{1}{s} & \dfrac{1+s^2}{s} \end{bmatrix}$$

$$\mathbf{z}(s) = \begin{bmatrix} 1 & 0 \\[3mm] 0 & \dfrac{s^2+s+1}{s+1} \end{bmatrix}$$

$$\mathbf{Z}(s) + \mathbf{z}(s) = \begin{bmatrix} \dfrac{1}{s}+1 & \dfrac{1}{s} \\[3mm] \dfrac{1}{s} & \dfrac{1+s^2}{s}+\dfrac{s^2+s+1}{s+1} \end{bmatrix}$$

$$= \begin{bmatrix} \dfrac{1+s}{s} & \dfrac{1}{s} \\[3mm] \dfrac{1}{s} & \dfrac{2s^3+2s^2+2s+1}{s(s+1)} \end{bmatrix}$$

$$[\mathbf{Z}(s) + \mathbf{z}(s)]^{-1} = \frac{s}{2(s^2 + s + 1)} \begin{bmatrix} \dfrac{2s^3 + 2s^2 + 2s + 1}{s(s+1)} & -\dfrac{1}{s} \\[3mm] -\dfrac{1}{s} & \dfrac{1+s}{s} \end{bmatrix}$$

$$= \frac{1}{2(s^2 + s + 1)} \begin{bmatrix} \dfrac{2s^3 + 2s^2 + 2s + 1}{s+1} & -1 \\[3mm] -1 & 1+s \end{bmatrix}$$

$$\mathbf{S}^I(s) = [\mathbf{Z}(s) + \mathbf{z}(s)]^{-1}[\mathbf{Z}(s) - \mathbf{z}(-s)]$$

$$= \frac{1}{2(s^2 + s + 1)} \begin{bmatrix} \dfrac{2s^3 + 2s^2 + 2s + 1}{s+1} & -1 \\[3mm] -1 & 1+s \end{bmatrix} \begin{bmatrix} \dfrac{1}{s} - 1 & \dfrac{1}{s} \\[3mm] \dfrac{1}{s} & \dfrac{1+s^2}{s} - \dfrac{s^2-s+1}{-s+1} \end{bmatrix}$$

$$= \frac{1}{2s(s^2 + s + 1)} \begin{bmatrix} \dfrac{-2s^4}{s+1} & \dfrac{-2s}{s^2-1} \\[3mm] 2s & \dfrac{2s^4}{s-1} \end{bmatrix}$$

$$\mathbf{r}(s) = \frac{1}{2}[\mathbf{z}(s) + \mathbf{z}_*(s)] = \frac{1}{2} \left\{ \begin{bmatrix} 1 & 0 \\[3mm] 0 & \dfrac{s^2+s+1}{s+1} \end{bmatrix} + \begin{bmatrix} 1 & 0 \\[3mm] 0 & \dfrac{s^2-s+1}{-s+1} \end{bmatrix} \right\}$$

$$= \begin{bmatrix} 1 & 0 \\[3mm] 0 & \dfrac{1}{1-s^2} \end{bmatrix} = \mathbf{h}\mathbf{h}_* = \begin{bmatrix} 1 & 0 \\[3mm] 0 & \dfrac{1}{1+s} \end{bmatrix} \begin{bmatrix} 1 & 0 \\[3mm] 0 & \dfrac{1}{1-s} \end{bmatrix}$$

$$S(s) = \mathbf{h}(s)S^I(s)\mathbf{h}_*^{-1}(s) = \begin{bmatrix} 1 & 0 \\ 0 & \dfrac{1}{1+s} \end{bmatrix} \dfrac{1}{2s(s^2+s+1)} \begin{bmatrix} \dfrac{-2s^4}{s+1} & \dfrac{-2s}{s^2-1} \\ 2s & \dfrac{2s^4}{s-1} \end{bmatrix} \begin{bmatrix} 1 & 0 \\ 0 & 1-s \end{bmatrix}$$

$$= \dfrac{1}{2s(s^2+s+1)} \begin{bmatrix} \dfrac{-2s^4}{s+1} & \dfrac{-2s}{s^2-1} \\ \dfrac{2s}{1+s} & \dfrac{2s^4}{s^2-1} \end{bmatrix} \begin{bmatrix} 1 & 0 \\ 0 & 1-s \end{bmatrix} = \dfrac{1}{2s(s^2+s+1)} \begin{bmatrix} \dfrac{-2s^4}{s+1} & \dfrac{-2s}{-(s+1)} \\ \dfrac{2s}{1+s} & \dfrac{2s^4}{-(s+1)} \end{bmatrix}$$

$$= \dfrac{1}{(s^2+s+1)} \begin{bmatrix} \dfrac{-s^3}{s+1} & \dfrac{1}{s+1} \\ \dfrac{1}{1+s} & \dfrac{s^3}{s+1} \end{bmatrix} = \dfrac{1}{(s+1)(s^2+s+1)} \begin{bmatrix} -s^3 & 1 \\ 1 & -s^3 \end{bmatrix}$$

Problem 4.25 *Repeat Example 4.4 for the steepness $1/k = 1.3$, everything else being the same.*

SOLUTION:

$$k = \dfrac{1}{1.3} = 0.7692308$$

$$RCf_c = 0.3183$$

From Appendix C, we have

$$r(y) = y^3 + 1.224882y^2 + 1.4631437y + 0.9734109$$

where $y = s/10^8$, since

$$\dfrac{2}{RC\omega_c} = 1 < C_2 = 1.224882$$

Case 2 applies and we choose $H_3 = 0.985$, yielding

$$\hat{\varepsilon} = \varepsilon\left(1 - H_3\right)^{-1/2} = 2.85212$$

We now proceed to compute $\hat{r}(y)$ as follows:

$$k' = \left(1 - k^2\right)^{1/2} = 0.638971029$$

$$k_1 = k^3\left[\frac{1 - \omega_1^2}{1 - k^2\omega_1^2}\right]^2 = 0.052636918$$

where

$$\omega_1 = \text{sn}\,(2K/3, k) = \text{sn}\,(1.29381, 0.7692308) = \sin 65.36° = 0.908945268$$

$$k'_1 = \left(1 - k_1^2\right)^{1/2} = 0.998613716$$

$$K = K(k) = 1.940717, \qquad K_1 = K(k_1) = 1.571886$$

$$K' = K(k') = 1.783308, \qquad K'_1 = K(k'_1) = 4.332944$$

$$\hat{a} = -j\frac{K}{nK_1}\,\text{sn}^{-1}\left(\frac{j}{\hat{\varepsilon}}, k_1\right) = \frac{K}{nK_1}F(19.3215,\ 0.99861) = 0.141483$$

$$\hat{y}_{p0} = -\,\text{tn}\,(\hat{a}, k') = -\,\text{tn}\,8.09° = -0.14214$$

$$\hat{y}_{p1} = j\,\text{sn}\,(\pm 1.29381 \pm j\hat{a}, k) = -4.195394 \times 10^{-2} \pm j\,0.9145205$$

$$\hat{r}(y) = y^3 + 0.226051y^2 + 0.850035y + 0.1191312$$

Set $\lambda = 1$, yielding

$$\hat{\rho}(y) = \frac{y^3 + 0.226051y^2 + 0.850035y + 0.119131}{y^3 + 1.224882y^2 + 1.463144y + 0.973411}$$

$$z_l(s) = \frac{100}{2y + 1}, \qquad A(s) = \frac{2y - 1}{2y + 1}, \qquad F(s) = \frac{-200}{(2y + 1)^2}$$

$$Z_{22}(s) = \frac{F(s)}{A(s) - \hat{\rho}(s)} - z_l(s) = \frac{42721.59966(y^2 + 0.613827y + 0.8552798)}{y^3 + 96.11420y^2 + 133.95167y + 467.2971}$$

which can also be obtained from (4.92) and (4.93), as follows:

$$d_2 - \hat{d}_2 = 1.224882 - 0.226051 = 0.998831$$

$$d_1 - \hat{d}_1 = 1.463144 - 0.850035 = 0.613109$$

$$d_0 - \hat{d}_0 = 0.973411 - 0.119131 = 0.85428$$

$$(d_3 + \hat{d}_3) - RC(d_2 - \hat{d}_2) = 2.338 \times 10^{-3}$$

$$(d_2 + \hat{d}_2) - RC(d_1 - \hat{d}_1) = 0.224715$$

$$(d_1 + \hat{d}_1) - RC(d_0 - \hat{d}_0) = 0.313179$$

$$d_0 + \hat{d}_0 = 1.092542$$

where $R = 100\ \Omega$ with normalized values equal to $R = 1\ \Omega$ and $C = 2$ F.

Problem 4.26 *Referring to the Darlington type-C load of Fig. 4.19, let $R_1 = 100\ \Omega$, $R_2 = 200\ \Omega$ and $C = 50$ pF. Design a third-order maximally-flat equalizer for this load. The cutoff frequency is 30 MHz.*

SOLUTION:

$$\omega = 2\pi f = 2\pi \times 30 \times 10^6 = 1.8849556 \times 10^8$$

giving the normalized load as

$$100\ \Omega\ \rightarrow\ 1$$

$$200\ \Omega\ \rightarrow\ 2$$

$$50 \times 10^{-12} \times 1.8849556 \times 10^8 \times 10^2 = 0.9424778\ \text{F}$$

$$\omega_c = 1$$

By (4-155) we have

$$r_l(s) = \frac{3 - 4C^2 s^2}{1 - 4C^2 s^2} = \frac{3 - 3.5530576 s^2}{1 - 3.5530576 s^2}$$

$$A(s) = \frac{2Cs - 1}{2Cs + 1} = \frac{1.8849556 s - 1}{1.8849556 s + 1}$$

$$z_I(s) = R_1 + \cfrac{1}{Cs + \cfrac{1}{R_2}} = 1 + \cfrac{1}{Cs + \cfrac{1}{2}} = \frac{3 + 2Cs}{1 + 2Cs} = \frac{3 + 1.8849556s}{1 + 1.8849556s}$$

$$F(s) = \frac{8C^2s^2 - 6}{(1 + 2Cs)^2} = \frac{7.1061152s^2 - 6}{4C^2s^2 + 4Cs + 1} = \frac{7.1061152s^2 - 6}{3.5530576s^2 + 3.7699112s + 1}$$

$$\frac{r_I(s)}{z_I(s)} = \frac{2C^2s^2 - \frac{3}{2}C}{(2Cs - 1)\left(s + \frac{3}{2}C\right)}$$

Thus, s_0 is a Class I zero of transmission of order 1 located at

$$s_0 = \sigma_0 = \frac{1}{2C}(1 + 2)^{1/2} = \frac{\sqrt{3}}{2C} = 0.91888$$

$$A_x = \rho_x, \quad x = 0$$

$$A_0 = A(s_0) = \frac{2C\dfrac{\sqrt{3}}{2C} - 1}{2C\dfrac{\sqrt{3}}{2C} + 1} = \frac{\sqrt{3} - 1}{\sqrt{3} + 1} = \frac{(\sqrt{3} - 1)^2}{2} = 2 - \sqrt{3}$$

For the third-order response, $n = 3$ and

$$G(\omega) = \frac{K_3}{1 + \omega^6}, \quad \omega \rightarrow -js$$

$$S_{22} = \alpha^3 \frac{p^3 + 2p^2 + 2p + 1}{s^3 + 2s^2 + 2s + 1} = \frac{s^3 + 2\alpha s^2 + 2\alpha^2 s + \alpha^3}{s^3 + 2s^2 + 2s + 1}$$

where $p = s/\alpha$, $\alpha^6 = 1 - K_3$, and

$$\rho(s) = S_{22}(s)$$

$$\rho_0 = \rho(s_0) = \frac{s_0^3 + 2\alpha s_0^2 + 2\alpha^2 s_0 + \alpha^3}{s_0^3 + 2s_0^2 + 2s_0 + 1} = A_0 = \frac{3}{2} - \sqrt{3} + \frac{1}{2} = 2 - \sqrt{3}$$

$$s_0^3 + 2\alpha s_0^2 + 2\alpha^2 s_0 + \alpha^3 = (2 - \sqrt{3})s_0^3 + 2(2 - \sqrt{3})s_0^2 + 2(2 - \sqrt{3})s_0 + (2 - \sqrt{3})$$

$$= (2s_0^3 - \sqrt{3}s_0^3) + 2(2 - \sqrt{3})s_0^2 + 2(2 - \sqrt{3})s_0 + (2 - \sqrt{3})$$

$$(2 - \sqrt{3} - 1)s_0^3 + (4 - 2\sqrt{3} - 2\alpha)s_0^2 + (4 - 2\sqrt{3} - 2\alpha^2)s_0 + (2 - \sqrt{3} - \alpha^3) = 0$$

$$\frac{3\sqrt{3}}{8C^3}\left(1 - \sqrt{3}\right) + \left(4 - 2\sqrt{3} - 2\alpha\right)\frac{3}{4C^2} + \left(4 - 2\sqrt{3} - 2\alpha^2\right)\frac{\sqrt{3}}{2C} + 2 - \sqrt{3} - \alpha^3 = 0$$

$$\frac{3\sqrt{3}}{8C^3} - \frac{9}{8C^3} + \frac{3}{C^2} - \frac{3\sqrt{3}}{2C^2} - \frac{3}{2C^2}\alpha + \frac{2\sqrt{3}}{C} - \frac{3}{C} - \frac{\sqrt{3}}{C}\alpha^2 + 2 - \sqrt{3} - \alpha^3$$

$$= \alpha^3 + \frac{\sqrt{3}}{C}\alpha^2 + \frac{3}{2C^2}\alpha + \left(\sqrt{3} - 2 + \frac{3}{C} - \frac{2\sqrt{3}}{C} + \frac{3\sqrt{3}}{2C^2} - \frac{3}{C^2} + \frac{9}{8C^3} - \frac{3\sqrt{3}}{8C^3}\right)$$

obtaining

$$C = 0.9424778$$

$$\alpha^3 + 1.837763\alpha^2 + 1.6886864\alpha - 0.64489592 = 0$$

$$\alpha > 0, \qquad \alpha = 0.2825$$

$$\rho(s) = \frac{s^3 + 0.565s^2 + 0.1596125s + 0.02254527}{s^3 + 2s^2 + 2s + 1}$$

$$= \frac{s^3 + 0.565s^2 + 0.1596125s + 0.02254527}{(s + 1)(s^2 + s + 1)}$$

$$1 - \alpha^6 = K_3, \qquad K_3 = 0.99949$$

$$Z_{22}(s) = \frac{F(s)}{A(s) - \rho(s)} - z_1(s)$$

$$= \frac{\dfrac{7.1061152s^2 - 6}{3.5530576s^2 + 3.7699112s + 1}}{\dfrac{1.8849556s - 1}{1.8849556s + 1} + \dfrac{s^3 + 0.505s^2 + 0.1596125s + 0.02254527}{(s + 1)(s^2 + s + 1)}} - \frac{3 + 1.8849556s}{1 + 1.8849556s}$$

$$= \frac{(7.1s^2 - 6)(s + 1)(s^2 + s + 1) - (3 + 1.885s)(0.7s^3 + 0.9s^2 - 0.32s - 1)}{(1.885s + 1)(0.7s^3 + 0.9s^2 - 0.32s - 1)}$$

$$= \frac{7.1s^4 + 19.41s^3 + 22.25s^2 + 13.46s + 3.232}{(1.885s + 1)(0.7s^2 + 1.544s + 1.1)}$$

$$Z_{22}(s) = \frac{3.7666s^3 + 8.3s^2 + 7.4s + 3.125}{0.7s^2 + 1.544s + 1.1}$$

$$= 5.38s + \cfrac{1}{0.47s + \cfrac{1}{1.35336s + 2.92}}$$

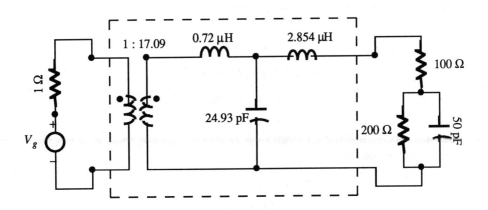

Problem 4.28 *Repeat Problem 4.26 for a third-order equiripple equalizer.*

SOLUTION:

$$A(s) = \frac{10^{-8}s - 1}{10^{-8}s + 1}$$

where

$$R_2C = 10^{-8}$$

$$\sigma_0 = \frac{1}{10^{-8}} \left(1 + \frac{200}{100}\right)^{1/2} = \sqrt{3} \times 10^8$$

$$A(\sigma_0) = \frac{\sqrt{3} - 1}{\sqrt{3} + 1} = 0.267949$$

$$y_0 = \frac{\sigma_0}{\omega_c} = 0.91888$$

For the Chebyshev response with $n = 3$ and $\varepsilon = 0.508847$, we have

$$p(y) = y^3 + 0.98834y^2 + 1.23841y + 0.49131$$

where $y = s/\omega_c$ and $\omega_c = 60\pi \times 10^6$ rad/s, giving

$$p(y_0) = 3.2396$$

$$\hat{p}_m(y) = y^3 + 0.75y$$

$$\hat{p}_m(y_0) = 1.465012$$

Since $A(\sigma_0)p(y_0) = 0.86805 < 1.465012 = \hat{p}_m(y_0)$, Case 1 applies. Thus, $K_3 = 1$ and we need to insert an open RHS zero

$$\sigma_1 = \sqrt{3} \times 10^8 \times \frac{1.465012 - 0.86805}{1.465012 + 0.86805} = 0.44318 \times 10^8$$

in $p(y)$, obtaining

$$\rho(y) = \frac{(y - 0.2351)(y^3 + 0.75y)}{(y + 0.2351)\left(y^3 + 0.98834y^2 + 1.23841y + 0.49131\right)}$$

$$Z_{22}(s) = \frac{376.99y^4 + 532.72y^3 + 470.55y^2 + 330.41y + 37.35}{0.7493y^3 + 1.0588y^2 + 0.5594y + 0.1257}$$

$$= 2.67 \times 10^{-6} s + \cfrac{1}{2.10 \times 10^{-11} s + \cfrac{1}{2.45 \times 10^{-6} s + \cfrac{1}{5.76 \times 10^{-11} s + \cfrac{1}{300}}}}$$

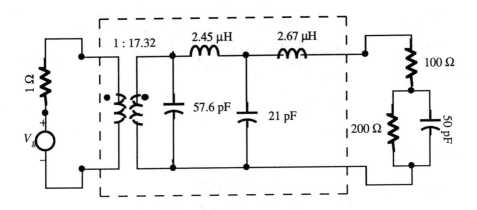

Problem 4.30 *Design an equalizer to match the load as indicated in (4.284) to a resistive generator and to achieve the third-order Chebyshev transducer power gain having a maximum attainable constant K_3. The passband tolerance is 1 dB and the normalized radian cutoff frequency $\omega_c = 1$.*

SOLUTION:

$$z_1(-s) = -s + \frac{1}{1-s}$$

$$r_1(s) = \frac{1}{1-s^2}$$

$$w(s) = \frac{r_1(s)}{z_1(s)} = \frac{1}{(1-s)(s^2+s+1)}$$

Thus, $s_0 = \infty$ is a Class IV zero of transmission of order $k = 3$. The coefficient constraints become

$$A_0 = \rho_0 \quad \rightarrow \quad A_0 = 1 = \rho_0 = 1$$

$$A_1 = \rho_1 \quad \rightarrow \quad A_1 = -2 = -0.98834 - 2\lambda$$

$$A_2 = \rho_2 \quad \rightarrow \quad A_2 = 2 = 0.488416 + 2 \times 0.98834\lambda + 2\lambda^2$$

$$\frac{F_2}{A_3 - \rho_3} \geq a_{-1} = 1 \quad \rightarrow \quad \frac{-2}{-2 - \rho_3} \geq 1$$

where

$$A(s) = \frac{s-1}{s+1} = 1 - \frac{2}{s} + \frac{2}{s^2} - \frac{2}{s^3} + \dots$$

$$F(s) = \frac{-2}{(1+s)^2} = 0 + 0 - \frac{2}{s^2} + \frac{4}{s^3} + \dots$$

and for $K_3 = 1$,

$$\rho(s)) = \eta(s)\hat{\rho}(s) = \frac{s-\lambda}{s+\lambda} \times \frac{s(s^2 + 0.75)}{s^3 + 0.98834s^2 + 1.2384s + 0.49131}$$

$$= \left(1 - \frac{2\lambda}{s} + \frac{2\lambda^2}{s^2} - \frac{2\lambda^3}{s^3} + \dots\right)\left(1 - \frac{0.98834}{s} + \frac{0.488416}{s^2} + \frac{0.249929}{s^3} + \dots\right)$$

$$= 1 + \frac{\rho_1}{s} + \frac{\rho_2}{s^2} + \frac{\rho_3}{s^3} + \dots$$

obtaining

$$\rho_1 = -0.98834 - 2\lambda$$

$$\rho_2 = 0.488416 + 2 \times 0.98834\lambda + 2\lambda^2$$

$$\rho_3 = 0.249920 - 2 \times 0.488416\lambda + 2(-0.98834)\lambda^2 - 2\lambda^3$$

Solving these equations yields

$$\lambda = 0.50583, \quad \rho_3 = -1$$

Thus, we have

$$\rho(s) = \frac{s - 0.50583}{s + 0.50583} \times \frac{s^3 + 0.75s}{s^3 + 0.98834s^2 + 1.2384s + 0.49131}$$

The equalizer back-end impedance is found to be

$$Z_{22}(s) = \frac{F(s)}{A(s) - \rho(s)} - z_1(s) = \frac{\dfrac{-2}{(1+s)^2}}{\dfrac{s-1}{s+1} - \rho(s)} - \frac{s^2 + s + 1}{s + 1}$$

$$= \frac{y^3 + 0.507y^2 + 1.24y + 0.257}{0.9912y^2 + 0.4898y + 0.2485}$$

$$= 1s + \cfrac{1}{1s + \cfrac{1}{3.98s + \cfrac{1}{0.967}}}$$

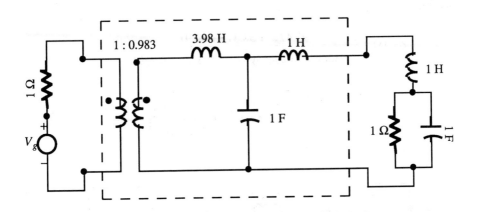

Problem 4.31 *Consider the same specifications as given in Example 4.4 except that the edge of the stopband starts at $60/\pi$ MHz and the ripple in the passband must not exceed 1 dB. Design this elliptic equalizer.*

SOLUTION:

$$\varepsilon = 0.508847$$

$$k = \frac{10^8}{1.2\times10^8} = 0.8333..., \qquad k_1 = 0.07967$$

$$K = 2.067255, \qquad K_1 = 1.5733$$

For $k = 0.83333$, $n = 3$, and $\varepsilon = 0.508847$, we have

$$r(y) = y^3 + 0.9747343y^2 + 1.230317y + 0.7291082$$

where $y = s/\omega_c$ and $\omega_c = 10^8$. Since

$$\frac{2}{RC\omega_c} = 1 > C_2 = 0.9747343$$

Case 1 applies and $H_3 = 1$.

$$\hat{\rho}(y) = \frac{y(y^2 + 0.847225)}{y^3 + 0.9747y^2 + 1.2303y + 0.7291}$$

yielding

$$Z_{22}(s) = \frac{97.47y^2 + 38.31y + 72.91}{0.0506y^3 + 0.20855y^2 + 0.619325y + 0.7291}$$

$$= \cfrac{1}{5.19\times10^{-12}s + \cfrac{1}{1.68\times10^{-6}s + \cfrac{1}{7.98\times10^{-11}s + \cfrac{1}{100}}}}$$

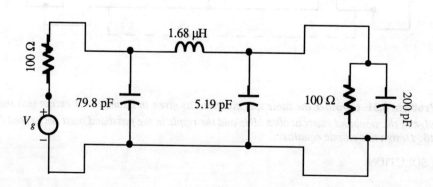

Problem 4.34 *Repeat the problem stated in Example 4.3 for n = 3.*

SOLUTION: Since

$$\frac{2 \sin (\pi/2n)}{RC\omega_c} = \sin \frac{\pi}{6} = 0.5 > \sinh a = \sinh\left(\frac{1}{3}\sinh^{-1}\frac{1}{\varepsilon}\right) = 0.49417$$

Case 1 applies. Thus, $K_3 = 1$ and

$$\hat{\rho}(s) = \frac{s^3 + 0.75s}{s^3 + 0.98834s^2 + 1.23841s + 0.49131}$$

$$Z_{22}(s) = \frac{0.98834s^2 + 0.48841s + 0.49131}{0.02332s^3 + 0.01152s^2 + 1.0058s + 0.49131}$$

$$= \cfrac{1}{2.3595\times10^{-2}s + \cfrac{1}{0.9941s + \cfrac{1}{2.024s + \cfrac{1}{1}}}}$$

Denormalizing the elements using magnitude-scaling by a factor 100 and frequency-scaling by 10^8 gives the final design.

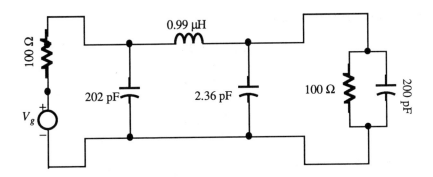

Problem 4.37 *Consider the same problem as stated in Example 4.6 except now that we wish to achieve the fourth-order Chebyshev transducer power gain with a passband tolerance of 1 dB. Plot the transducer power gain of your realization as a function of ω.*

SOLUTION: Choose $K_4 = 1$. Then

$$\hat{\rho}(s) = \frac{s^4 + s^2 + 0.125}{s^4 + 0.9528s^3 + 1.4539s^2 + 0.7426s + 0.275}$$

$$\eta(s) = \frac{s - \lambda}{s + \lambda}, \qquad \lambda > 0$$

$$\rho(s) = \eta(s)\hat{\rho}(s) = \left(1 + \frac{\eta_1}{s} + \frac{\eta_2}{s^2} + \frac{\eta_3}{s^3} + \cdots\right)\left(1 + \frac{\hat{\rho}_1}{s} + \frac{\hat{\rho}_2}{s^2} + \frac{\hat{\rho}_3}{s^3} + \cdots\right)$$

$$= 1 + \frac{\rho_1}{s} + \frac{\rho_2}{s^2} + \frac{\rho_3}{s^3} + \cdots$$

where

$$\rho_1 = \hat{\rho}_1 + \eta_1 = -0.95281 - 2\lambda$$

$$\rho_2 = \hat{\rho}_2 + \eta_1\hat{\rho}_1 + \eta_2 = 0.453917 + 2\times0.95281\lambda + 2\lambda^2$$

$$\rho_3 = 0.210202 - 2\times0.453917\lambda - 2\times0.95281\lambda^2 - 2\lambda^3$$

The coefficient constraints become

$$\rho_1 = A_1 = 2$$

$$\rho_2 = A_2 = -2$$

$$\frac{-2}{-2 - \rho_3} \geq 1$$

Solving the above equations yields

$$\lambda = 0.523595, \qquad \rho_3 = -1.07465$$

$$\rho(s) = \frac{s - 0.523595}{s + 0.523595} \times \frac{s^4 + s^2 + 0.125}{s^4 + 0.9528s^3 + 1.4539s^2 + 0.7426s + 0.2756}$$

$$Z_{22}(s) = \dfrac{\dfrac{-2}{(1+s)^2}}{\dfrac{s-1}{s+1} - \rho(s)} - \dfrac{s^2+s+1}{s+1}$$

$$= \dfrac{1.0746s^5 + 1.5866s^4 + 1.9597s^3 + 1.9083s^2 + 0.6703s + 0.2098}{(s+1)(0.9253s^3 + 0.4408s^2 + 0.5796s + 0.0789)}$$

$$= \dfrac{1.0746s^4 + 0.512s^3 + 1.4477s^2 + 0.4606s + 0.2097}{0.9253s^3 + 0.4408s^2 + 0.5797s + 0.0789}$$

or

$$Z_{22}(s) = 1.1614s + \cfrac{1}{1.1948s + \cfrac{1}{2.3529s + \cfrac{1}{1.5697s + \cfrac{1}{2.6578}}}}$$

Denormalizing the elements yields the following design:

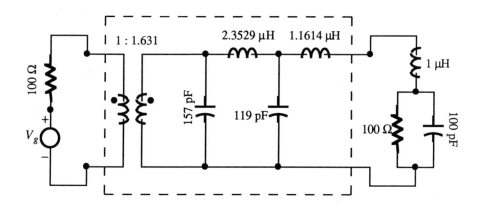

Problem 4.38 *Using the second-order Chebyshev transducer power gain with passband tolerance of 1 dB, show that the load given in Example 4.7 can always be matched to a resistive generator for any ω_c.*

106

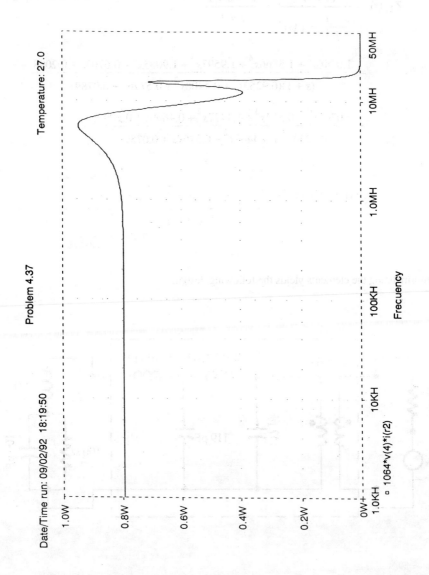

SOLUTION:

$$\sigma_0 = \frac{1}{R_2 C}\left(1 + \frac{R_2}{R_1}\right)^{1/2} = 10^8, \qquad y_0 = \frac{\sigma_0}{\omega_c} = \frac{10^8}{\omega_c}$$

$$A(\sigma_0) = \frac{R_2 C \sigma_0 - 1}{R_2 C \sigma_0 + 1} = \frac{1}{3}$$

For the second-order Chebyshev response, we have

$$\rho(y_0) = y_0^2 + 1.0977 y_0 + 1.1025$$

$$\widehat{\rho}_m(y_0) = y_0^2 + 0.5$$

Consider the following inequality:

$$A(\sigma_0)\rho(y_0) < \widehat{\rho}_m(y_0)$$

or

$$\frac{1}{3}\left(y_0^2 + 1.0977 y_0 + 1.1025\right) < y_0^2 + 0.5$$

Simplifying it yields

$$y_0^2 - 0.54885 y_0 + 0.19875 > 0$$

Since

$$(-0.54885)^2 - 4 \times 0.19875 = -0.49376 < 0$$

the above two inequalities are always satisfied. This means that for any ω_c, the load given in Example 4.7 can always be matched to a resistive generator by choosing

$$\rho(y) = \frac{y - \sigma_1}{y + \sigma_1} \times \frac{y^2 + 0.5}{y^2 + 1.0977y + 1.1025}$$

for Case 1, where

$$\sigma_1 = 10^8 \times \frac{(y_0^2 + 0.5) - \frac{1}{3}\left(y_0^2 + 1.0977y_0 + 1.1025\right)}{(y_0^2 + 0.5) + \frac{1}{3}\left(y_0^2 + 1.0977y_0 + 1.1025\right)}$$

$$y_0 = \frac{10^8}{\omega_c}$$

Problem 4.40 *Repeat the problem stated in Example 4.8 for n = 3.*

SOLUTION:

Step 1. $\qquad\qquad\qquad\qquad 0 \le K_3 \le 1$

Step 2. Frequency-scale the capacitance C by a factor of

$$a = \frac{1}{0.5 \times 10^8} \qquad \rightarrow \qquad C = \frac{10}{3} \times 10^{-3}$$

$$z_I(s) = R_1 + \frac{R_2}{1 + R_2 Cs} - \frac{(R_1 + R_2) + R_1 R_2 Cs}{1 + R_2 Cs}$$

$$= \frac{400 + 100s}{1 + s} = 100\,\frac{s + 4}{s + 1} \quad \leftarrow \quad \text{normalized (scaled)}$$

$$r_I(s) = \frac{1}{2}\left[z_I(s) + z_{I*}(s)\right] = \frac{(R_1 + R_2) - R_1 R_2^2 C^2 s^2}{1 - R_2^2 C^2 s^2} = 100\,\frac{4 - s^2}{1 - s^2}$$

$$A(s) = \frac{R_2 Cs - 1}{R_2 Cs + 1} = \frac{300\left(\frac{10}{3} \times 10^{-3}\right)s - 1}{300\left(\frac{10}{3} \times 10^{-3}\right)s + 1} = \frac{s - 1}{s + 1} \qquad \text{(scaled)}$$

$$F(s) = 2r(s)A(s) = \frac{2R_1 R_2^2 C^2 s^2 - 2(R_1 + R_2)}{(1 + R_2 Cs)^2} = 200\,\frac{s^2 - 4}{(s + 1)^2}$$

Step 3.

$$w(s) = \frac{r_l(s)}{z_l(s)} = \frac{R_2 C s^2 - (R_1 + R_2)/R_1 R_2 C}{(R_2 C s - 1)\left[s + (R_1 + R_2)/R_1 R_2 C\right]}$$

$\Rightarrow z_l(s)$ has a Class I zero of transmission of order 1 located at

$$s_0 = \sigma_0 = \frac{1}{R_2 C}\left(1 + \frac{R_2}{R_1}\right)^{1/2} = 10^8$$

The scaled zero is found to be

$$y_0 = \frac{\sigma_0}{0.5 \times 10^8} = 2$$

$$A(\sigma_0) = \frac{1}{3} = 0.3333...$$

for $n = 3 \quad \Rightarrow \quad q(s) = 1 + 2s + 2s^2 + s^3$

$$A(\sigma_0)q\left(\frac{1}{y_0}\right) = \frac{1}{3}(1 + 1 + 0.5 + 0.125) = 0.875 < 1$$

Case 1 applies, indicating that $K_3 = 1$ is attainable by the insertion of an open RHS zero in $\rho(s)$.

$$\Rightarrow \quad \sigma_1 = \sigma_0 \frac{1 - A(\sigma_0)q\left(\frac{\omega_c}{\sigma_0}\right)}{1 + A(\sigma_0)q\left(\frac{\omega_c}{\sigma_0}\right)} > 0$$

to

$$\sigma_0 = \frac{1 - 0.875}{1 + 0.875} 10^8 = 6.6667 \times 10^6$$

Using (4.137) and (4.158) obtains

$$\rho(s) = \pm \eta(s)\hat{\rho}(s) = \pm \eta(s)\left(1 - K_3\right)^{1/2}\frac{q(x)}{q(y)}$$

where

$$x = \left(1 - K_3\right)^{-1/6} \frac{s}{\omega_s} = \left(1 - K_3\right)^{-1/6} y$$

$$y = \frac{s}{\omega_s} = \frac{s}{0.5 \times 10^6} = 2 \times 10^{-8} s$$

$$\Rightarrow \quad \rho(y) = \frac{(y - 0.1333)\left(1 - K_3\right)^{1/2} q\left[\left(1 - K_3\right)^{-1/6} y\right]}{(y + 0.13333) q(y)}$$

$$= \frac{(y - 0.13333) y^3}{(y + 0.13333)\left(y^3 + 2y^2 + 2y + 1\right)}$$

where

$$y = \frac{s}{\omega_c} = 2 \times 10^{-8} s$$

the normalized complex frequency.

$$Z_{22}(s) = Z_{22}(y) = \frac{F(y)}{A(y) - \rho(y)} - z_1(y)$$

$$= \frac{200 \dfrac{y^2 - 4}{(y + 1)^2}}{\dfrac{y - 1}{y + 1} - \dfrac{(y - 0.13333) y^3}{(y - 0.13333)\left(y^3 + 2y^2 + 2y + 1\right)}} - 100 \frac{y + 4}{y + 1}$$

$$= 750 \frac{y^4 + 3y^3 + 2.733 y^2 + 1.7y + 0.134}{y^3 + 2y^2 + 2.25y + 0.25}$$

where

$$y = 2 \times 10^{-8} s$$

Problem 4.42 *Repeat the problem stated in Example 4.9 for n = 3.*

SOLUTION: For $n = 3$ and $\varepsilon = 0.508847$, the required functions for the Chebyshev response are found to be

$$P(s) = s^3 + 0.98834s^2 + 1.23841s + 0.49131$$

$$\hat{P}_m(y) = y^3 + 0.75y$$

$$\sigma_0 = \frac{1}{2\times10^{-8}}(1+3)^{1/2} = 10^8, \qquad y_0 = 1$$

$$A(s) = \frac{2\times10^{-8}s - 1}{2\times10^{-8}s + 1}, \qquad A(\sigma_0) = \frac{1}{3}$$

$$P(y_0) = 3.71806, \qquad \hat{P}_m(y_0) = 1.75$$

where $y = s/10^8$. Since

$$A(\sigma_0)P(y_0) = 1.23935 < 1.75 = \hat{P}_m(y_0)$$

Case 1 applies. Thus, $K_3 = 1$ and we insert an open RHS zero

$$\sigma_1 = 10^8 \frac{1.75 - 1.23935}{1.75 + 1.23935} = 0.17082\times10^8$$

in $\rho(y)$ and obtain

$$\rho(y) = \frac{(y - 0.17082)(y^3 + 0.75y)}{(y + 0.17082)\left(y^3 + 0.98834y^2 + 1.23841y + 0.49131\right)}$$

and

$$Z_{22}(s) = \frac{400y^4 + 597.67y^3 + 497.13y^2 + 349.18y + 33.58}{0.66y^3 + 0.986y^2 + 0.491y + 0.084}$$

$$= 606.1y + \cfrac{1}{3.307\times10^{-3}y + \cfrac{1}{525.2y + \cfrac{1}{1.1314\times10^{-2}y + \cfrac{1}{400}}}}$$

$$= 6.06 \times 10^{-6} s + \cfrac{1}{33.07 \times 10^{-12} s + \cfrac{1}{5.252 \times 10^{-6} s + \cfrac{1}{113.14 \times 10^{-12} s + \cfrac{1}{400}}}}$$

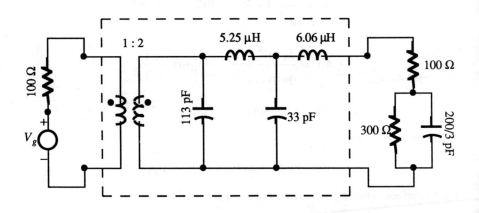

Problem 4.45 *Repeat the problem stated in Example 4.10 for n = 3.*

SOLUTION: For $n = 3$ and $\varepsilon = 0.508847$, the required functions for the Chebyshev response are found to be

$$P(s) = s^3 + 0.98834s^2 + 1.23841s + 0.49131$$

$$\widehat{P}_m(y) = y^3 + 0.75y$$

$$\sigma_0 = \frac{1}{2 \times 10^{-8}} \left(1 + \frac{300}{5} \right)^{1/2} = 3.90512 \times 10^8$$

$$y_0 = \frac{\sigma_0}{\omega_c} = 3.90512$$

$$A(s) = \frac{2\times10^{-8}s - 1}{2\times10^{-8}s + 1}, \qquad A(\sigma_0) = \frac{2\times3.90512 - 1}{2\times3.90512 + 1} = 0.77299$$

$$P(y_0) = 79.95279, \qquad \hat{P}_m(y_0) = 62.48199$$

where $y = s/10^8$. Since

$$A(\sigma_0)P(y_0) = 61.80084 < 62.48199 = \hat{P}_m(y_0)$$

Case 1 applies. Thus, $K_3 = 1$ and we need to insert an open RHS zero

$$\sigma_1 = 3.90512\times10^8 \frac{62.48199 - 61.80284}{62.48199 + 61.80284} = 0.02134\times10^8$$

in $\rho(y)$ and obtain

$$\rho(y) = \frac{(y - 0.02134)(y^3 + 0.75y)}{(y + 0.02134)\left(y^3 + 0.98834y^2 + 1.23841y + 0.49131\right)}$$

$$Z_{22}(s) = \frac{20y^5 + 9.8834y^4 - 294.366y^3 - 150.38y^2 - 162.69y - 3.1964}{0.06204y^4 + 0.03066y^3 - 0.942y^2 - 0.4808y - 0.0105}$$

$$= 322.373y + \cfrac{1}{0.00665877y + \cfrac{1}{78.7174y + \cfrac{1}{\cfrac{40.79178279y + 0.8184752683}{0.118361y + 0.00268874}}}}$$

which can be realized by an LC ladder network terminating in the impedance

$$z(s) = \frac{40.79178279y + 0.8184752683}{0.118361y + 0.00268874}$$

the even part of which is found to be

$$r(s) = \frac{4.8281559y^2 - 0.0022006665}{0.0140093y^2 - 7.229323\times10^{-6}}$$

Thus, $\sigma_0 = 0.0213494$ is an open RHS zero of $r(s)$. Using Youla's cascade synthesis formula, we can realize this zero by a Darlington type-C section. The final realization of the matching network is shown below.

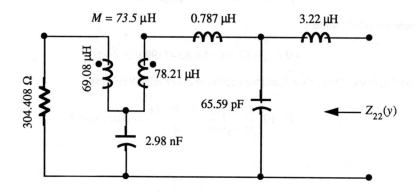

Problem 4.46 *Equalize the load impedance*

$$z_l(s) = \frac{s^2 + s + 2}{2s^2 + s + 2}$$

to a resistive generator and to achieve a truly-flat transducer gain over the entire real-frequency axis. Obtain the maximum attainable constant transducer power gain ζ, and compute the equalizer back-end impedance.

SOLUTION: The desired functions are found to be

$$r_l(s) = \frac{2s^4 + 5s^2 + 4}{(2s^2 + s + 2)(2s^2 - s + 2)}$$

$$w(s) = \frac{r_l(s)}{z_l(s)} = \frac{2s^4 + 5s^2 + 4}{(2s^2 - s + 2)(2s^2 + s + 2)}$$

The load impedance $z_l(s)$ has a pair of Class I zeros of transmission of order 1 located at

$$s_0,\ \bar{s}_0 = 0.2865 \pm j1.1542 = \sigma_0 \pm j\omega_0$$

obtaining

$$A(s) = \frac{2s^2 - s + 2}{2s^2 + s + 2}, \qquad A(s_0) = 0.0947 + j0.3094 = a_1 + ja_2$$

From (4.237) and (4.240)

$$\gamma_{min} = 0.2865 \left| \frac{0.30939}{1.1542} \right| + \left[\left(\frac{0.2865 \times 0.3094}{1.1542} \right)^2 + 0.3236^2 \right]^{1/2} = 0.40935$$

$$L = \frac{2 \times 0.40935 \times 0.3094}{1.1542 \left[(0.4094 - 0.0947)^2 + 0.3094^2 \right]} = 1.127$$

$$\rho(s) = \gamma_{min} \frac{Ls - 1}{Ls + 1} = 0.4094 \frac{1.127s - 1}{1.127s + 1}$$

$$Z_{22}(s) = \frac{F(s)}{A(s) - \rho(s)} - z_1(s) = \frac{\dfrac{2(2s^4 + 5s^2 + 4)}{(2s^2 + s + 2)^2}}{\dfrac{2s^2 - s + 2}{2s^2 + s + 2} - \gamma_{min}\dfrac{Ls - 1}{Ls + 1}} - \frac{s^2 + s + 2}{2s^2 + s + 2}$$

$$= \frac{3.1767s^5 + 1.4383s^4 + 6.6363s^3 + 5.9798s^2 + 6.7159s + 6.3626}{(2s^2 + s + 2)\left(1.3313s^3 + 1.2304s^2 + 0.7407s + 0.8187\right)}$$

Problem 4.48 *Repeat Problem 4.46 for the load impedance*

$$z_l(s) = \frac{5s^2 + 3s + 4}{s^2 + 2s + 2}$$

SOLUTION: Let

$$G(\omega^2) = K, \qquad 0 \le K \le 1$$

Then the even part of $z_2(s)$ is given by

$$r_2(s) = \frac{1}{2}\left[z_1(s) + z_1(-s) \right] = \frac{1}{2}\left[\frac{(5s^2 + 4) + 3s}{(s^2 + 2) + 2s} + \frac{(5s^2 + 4) - 3s}{(s^2 + 2) - 2s} \right]$$

$$= \frac{(5s^2 + 4)(s^2 + 2) - 6s^2}{(s^2 + 2s + 2)(s^2 - 2s + 2)} = \frac{5s^4 + 8s^2 + 8}{(s^2 + 2s + 2)(s^2 - 2s + 2)}$$

$$s^2 - 2s + 2 = (s - 1 + j)(s - 1 - j) = 0$$

The open RHS zeros of $z_1(-s)$ are found to be $1 \pm j$.

$$A(s) = \frac{s^2 - 2s + 2}{s^2 + 2s + 2}$$

$$\frac{r_2(s)}{z_1(s)} = \frac{5s^4 + 8s^2 + 8}{(s^2 - 2s + 2)(5s^2 + 3s + 4)}$$

$$5s^4 + 8s^2 + 8 = 5(s^4 + 1.6s^2 + 1.6) = 0$$

$$s^4 + 1.6s^2 + 1.6 = (s^2 + as^2 + \sqrt{1.6})(s^2 - as^2 + \sqrt{1.6}) = 0$$

$$(s^2 + \sqrt{1.6})^2 - a^2 s^2 = 0$$

$$s^4 + (2\sqrt{1.6} - a^2)s^2 + 1.6 = 0$$

$$2\sqrt{1.6} - a^2 = 1.6$$

giving

$$a = \pm 0.96427$$

$$\rightarrow \quad s_0 = \pm(0.48214 \pm j1.0161) = \sigma_0 + j\omega_0$$

The zeros of transmission are located at

$$s_0, \; \bar{s}_0 = +0.48214 \pm j1.0161 \quad \text{(RHS zeros)}$$

Both zeros belong to Class I zeros of transmission of order $k = 1$. Thus, we require that

$$A_0 = \rho_0$$

$$A(s_0) = \frac{(0.48214 + j1.0161)^2 - 2(0.48214 + j1.0161) + 2}{(0.48214 + j1.0161)^2 + 2(0.48214 + j1.0161) + 2}$$

$$= \frac{0.23568 - j1.0524}{2.1648 + j3.012} = -0.19331 - j0.21718 = a_1 + ja_2$$

Thus, Subcase 2 of Case 2 applies. From (4.237), we have

$$\gamma_{min} = \sigma_0 \frac{|a_2|}{|\omega_0|} + \left[\left(\frac{\sigma_0 a_2}{\omega_0} \right)^2 + |A(s_0)|^2 \right]^{\frac{1}{2}}$$

$$= 0.48214 \left| \frac{0.21718}{1.0161} \right| + \left[\left(\frac{0.48214 \times 0.21718}{1.0161} \right)^2 + 0.19331^2 + 0.21718^2 \right]^{\frac{1}{2}}$$

$$= 0.41153$$

$$K = 1 - \gamma_{min}^2 = 0.83065$$

Since

$$\frac{a_2}{\omega_0} < 0$$

choose from (4.241)

$$C = \frac{|\omega_0| \left[(\gamma_{min} - a_1)^2 + a_2^2 \right]}{2\gamma_{min}|a_2||s_0|^2} = \frac{1.0161 \left[(0.41153 + 0.19331)^2 + 0.21718^2 \right]}{2 \times 0.41153 \times 0.21718 \times (0.48214^2 + 1.0161^2)}$$

$$= 1.8560$$

$$\rho(s) = \gamma_{min} \frac{1 - Cs}{1 + Cs} = 0.41153 \frac{1 - 1.856s}{1 + 1.856s} = -0.41152 \frac{s - 0.5388}{s + 0.5388}$$

$$F(s) = 2 r_l(s) A(s) = 2 \frac{5s^4 + 8s^2 + 8}{(s^2 + 2s + 2)(s^2 - 2s + 2)} \times \frac{s^2 - 2s + 2}{s^2 + 2s + 2}$$

$$Z_{22}(s) = \frac{F(s)}{A(s) - \rho(s)} - z_l(s) = \frac{\dfrac{2\left(5s^4 + 8s^2 + 8\right)}{\left(s^2 + 2s + 2\right)^2}}{\dfrac{s^2 - 2s + 2}{s^2 + 2s + 2} + \dfrac{0.41152\,(s - 0.5388)}{s + 0.5388}} - \frac{5s^2 + 3s + 4}{s^2 + 2s + 2}$$

$$= \frac{2\left(5s^4 + 8s^2 + 8\right)(s + 0.5388)}{\left(s^2 + 2s + 2\right)\left(1.4115s^3 - 0.85989s^2 + 1.302s + 0.63415\right)} - \frac{5s^2 + 3s + 4}{s^2 + 2s + 2}$$

$A(s) - \rho(s) = 0$ at the two zeros of transmission s_0 and $\bar{s}_0$. $F(s)$ also contains these zeros, since its numerator is the same as $r_l(s)/z_l(s)$. Thus, both the numerator and denominator of the first function contain the term

$$\left(s - s_0\right)\left(s - \bar{s}_0\right) = s^2 - 0.86428s + 1.2649$$

which may be divided out, yielding

$$Z_{22}(s) = \frac{2 \times 5 \left(s^2 + 0.96428s + 1.2649\right)(s + 0.5388)}{\left(s^2 + 2s + 2\right)(1.4115s + 0.50119)} - \frac{5s^2 + 3s + 4}{s^2 + 2s + 2}$$

$$- \frac{2.9425s^3 + 8.2904s^2 + 10.695s + 4.8105}{\left(s^2 + 2s + 2\right)(1.4115s + 0.50119)}$$

Dividing $\left(s^2 + 2s + 2\right)$ out gives

$$Z_{22}(s) = \frac{2.9425s + 2.4054}{1.4115s + 0.50119} = 2.0847\,\frac{s + 0.81747}{s + 0.35508}$$

Without using formulas (4.236) and (4.241), we apply the constraint

$$\tilde{A}_0 = \tilde{\rho}_0$$

obtaining

$$\tilde{A}_0 = A(s_0) = A(0.48214 + j1.0161) = -0.19331 - j0.21718 \neq \tilde{\rho}_0 = \pm\sqrt{1 - K}$$

$$\rightarrow \quad \rho = \eta\hat{\rho} = \pm\frac{s - \sigma_1}{s + \sigma_1}\sqrt{1 - K}$$

$$A(s_0) = \rho(s_0)$$

$$\rightarrow \quad \left(\underline{-}\, 0.19331 - j\overset{.}{0}.21718\right)\left(s_0 + \sigma_1\right) = \pm\left(s_0 - \sigma_1\right)\sqrt{1 - K}$$

We must choose the "−" sign. Equating the real and imaginary parts obtains

$$-j0.30113 - j0.21718\sigma_1 = -j1.016\sqrt{1 - K}$$

$$0.12747 - 0.19331\sigma_1 = -\left(0.48214 - \sigma_1\right)\sqrt{1 - K}$$

yielding

$$\sqrt{1 - K} = 0.29636 + 0.21374\sigma_1$$

Combining these results in

$$\sigma_1^2 + 1.80886\sigma_1 - 1.2649 = 0$$

$$\sigma_1 = 0.5388 \quad \text{or} \quad -2.3476$$

From

$$\gamma_{min} = \sqrt{1 - K} = 0.41153$$

$$K = 1 - 0.41152^2 = 0.83065$$

we arrive at

$$\rho(s) = -0.41152\,\frac{s - 0.5388}{s + 0.5388}$$

Problem 4.49 *Determine the maximum dc gain K_3 and the corresponding equalizer back-end impedance $Z_{22}(s)$ of Problem 4.23 for the load*

$$z_l(s) = 3s + \frac{1}{s + 1}$$

[Hint. We must insert open RHS zeros in $\rho(s)$.]

SOLUTION:

1) If $K_3 = 1$, $G(\omega^2) \leq 1$ for all ω.

2)

$$r_l(s) = \frac{1}{2}\left[z_l(s) + z_l(-s)\right]^{\cdot} = \frac{1}{1-s^2}$$

$$z_l(-s) = \frac{3s^2 - 3s + 1}{-s + 1}$$

$$A(s) = \frac{s-1}{s+1}$$

$$F(s) = 2r_l(s)A(s) = \frac{2}{1-s^2} \times \frac{s-1}{s+1} = \frac{-2}{(s+1)^2}$$

3) The zeros of transmission are obtained from

$$\frac{r_l(s)}{z_l(s)} = \frac{\dfrac{1}{1-s^2}}{\dfrac{3s^2+3s+1}{s+1}} = \frac{1}{(1-s)(3s^2+3s+1)}$$

Thus, $s_0 = \sigma_0 + j\omega_0 = 0 + j\infty$ is a Class IV zero of transmission of order $k = 3$.

4)

$$\rho(s)\rho(-s) = 1 - G(-s^2) = 1 - \frac{K_3}{1-s^6} = \frac{1-K_3-s^6}{1-s^6}$$

$$1 - K_3 = \alpha^6$$

Thus, $\rho(s)$ in general is of the form

$$\rho(s) = \eta\hat{\rho}(s)$$

$$\rho(s) = \frac{s-\sigma}{s+\sigma} \times \frac{s^3 + 2\alpha s^3 + 2\alpha^2 s + \alpha^3}{s^3 + 2s^2 + 2s + 1}$$

5)

$$A(s) = \frac{s-1}{s+1} = A_0 + \frac{A_1}{s} + \frac{A_2}{s^2} + \frac{A_3}{s^3} + \ldots = 1 - \frac{2}{s} + \frac{2}{s^2} - \frac{2}{s^3} + \ldots$$

$$F(s) = \frac{-2}{(s+1)^2} = F_0 + \frac{F_1}{s} + \frac{F_2}{s^2} + \frac{F_3}{s^3} + \ldots = 0 + \frac{0}{s} - \frac{2}{s^2} + \frac{4}{s^3} + \ldots$$

$$\rho(s) = 1 + \frac{2\alpha - 2 - 2\sigma}{s} + \frac{2\alpha^2 + 2\sigma^2 - 4\alpha\sigma + 4\sigma - 4\alpha + 2}{s^2}$$

$$+ \frac{\alpha^3 - 4\alpha^2\sigma + 4\alpha\sigma^2 - 2\sigma^3 - 4\sigma^2 + 8\alpha\sigma - 4\alpha^2 - 4\alpha - 4\sigma - 1}{s^3} + \ldots$$

6) If $K_3 = 1$, then

$$\alpha^6 = 1 - K_3 = 1 - 1 = 0 \quad \Rightarrow \quad \alpha = 0$$

The condition

$$A_0 = \rho_0 = 1$$

is satisfied.

$$A_1 = -2 = \rho_1 = (2\alpha - 2 - 2\sigma)|_{\alpha=0} = -2 - 2\sigma$$

$$-2 = -2 - 2\sigma \quad \Rightarrow \quad \sigma = 0$$

$$A_2 = 2 = \rho_2 = 2\alpha^2 + 2\sigma^2 - 4\alpha\sigma + 4\sigma - 4\alpha + 2 \,|_{\alpha=0,\ \sigma=0} = 2$$

$$\frac{F_2}{A_3 - \rho_3} = \frac{-2}{-2 - (-1)} = \frac{-2}{-1} = 2 = L_\alpha$$

With $K_n = 1$, the match is possible without the insertion of the open RHS zeros in $\rho(s)$ if $L_\alpha \geq 3$. Since

$$\frac{F_2}{A_3 - \rho_3} = 2$$

is not greater than or equal to 3, we must find a positive real σ such that

$$\rho(s) = \frac{s - \sigma}{s + \sigma} - \frac{s^3 + 2\alpha s^2 + 2\alpha^2 s + \alpha^3}{s^3 + 2s^2 + 2s + 1} = \rho_0 + \frac{\rho_1}{s} + \frac{\rho_2}{s^2} + \frac{\rho_3}{s^3}$$

$$\rho_0 = 1 = A_0 = 1$$

$$A_1 = -2 = \rho_1 = -2\alpha - 2 - 2\sigma \quad \Rightarrow \quad \alpha = \sigma$$

If $\alpha = \sigma$, then

$$A_2 = \rho_2$$

Considering $\alpha = \sigma$, we find ρ_3 to be

$$\rho_3 = -\alpha^3 - 1$$

$$\frac{F_2}{A_3 - \rho_3} = \frac{-2}{-2 + \alpha^3 + 1} \geq 3$$

We must determine the minimum α satisfying the above relations. This minimum α will result in the maximum gain K_3.

$$\frac{-2}{\alpha^3 - 1} = \frac{2}{1 - \alpha^3} \geq 3, \qquad \alpha^3 \geq \frac{1}{3}$$

obtaining

$$\alpha = 0.69336127, \qquad \sigma = 0.69336127$$

$$K_3 = 1 - \alpha^6 = 0.88888889$$

7)

$$Z_{22}(s) = \frac{F(s)}{A(s) - \rho(s)} - z_1(s)$$

We solve this in terms of α and obtain

$$Z_{22}(s) = \frac{\dfrac{-2}{(s+1)^2}}{\dfrac{s-1}{s+1} - \dfrac{(s-\alpha)(s^3 + 2\alpha s^2 + 2\alpha^2 s + \alpha^3)}{(s+\alpha)(s^3 + 2s^2 + 2s + 1)}} - \frac{3s^2 + 3s + 1}{s+1}$$

$$= \frac{\dfrac{-2}{(s+1)^2}}{\dfrac{s-1}{s+1} - \dfrac{(s-\alpha)(s^2+\alpha s + \alpha^2)}{(s+1)(s^2+s+1)}} - \frac{3s^2+3s+1}{s+1}$$

$$= \frac{2}{s+1}$$

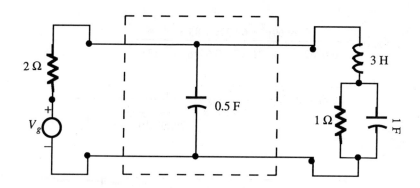

Problem 4.50 *Repeat the problem stated in Example 4.4 for a passband ripple of 0.43 dB, everything else being the same.*

SOLUTION:

$$\varepsilon = \left(10^{0.043} - 1\right)^{1/2} = 0.322612, \qquad K = 1.8628$$

$$RCf_c = 0.3183, \quad k_1 = 0.03754, \quad K_1 = 1.57135$$

For $k = 0.7142857$, $n = 3$, and $\varepsilon = 0.322612$

$$r(y) = y^3 + 1.2773y^2 + 1.5428y + 0.9945$$

where $y = s/10^8$. Since

$$\frac{2}{RC\omega_c} = 1 < C_2 = 1.2773$$

Case 2 applies, and we choose

$$H_3 = 0.98, \qquad \hat{\varepsilon} = 2.2812$$

$$\hat{r}(y) = y^3 + 0.2804y^2 + 0.8496y + 0.1407$$

Thus, we have

$$\hat{\rho}(y) = \frac{y^3 + 0.2804y^2 + 0.8496y + 0.1407}{y^3 + 1.2773y^2 + 1.5428y + 0.9945}$$

$$Z_{22}(s) = \frac{99.69y^2 + 69.32y + 85.38}{0.0062y^3 + 0.1713y^2 + 0.6848y + 1.1352}$$

$$= \cfrac{1}{0.62 \times 10^{-12}s + \cfrac{1}{1.47 \times 10^{-6}s + \cfrac{1}{79.58 \times 10^{-12}s + \cfrac{1}{75.21}}}}$$

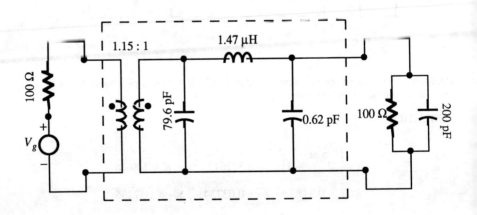

Problem 4.53 *Repeat Problem 4.52 for the Chebyshev response described in §4.2.*

SOLUTION: A general factorization of (4.48a) is given by

$$\rho(s) = \left(1 - K_n\right)^{1/2} \frac{\hat{q}(y)}{q(y)}$$

As we know, $q(y)$ is a Hurwitz polynomial, and $\hat{q}(y)$ may possess zeros in the LHS and the RHS. Let $-y_j$ and y_j be zeros of $\hat{q}(y)$ in the LHS and RHS, respectively, and

$$\rho(s) = \left(1 - K_n\right)^{1/2} \frac{\hat{q}(y)}{q(y)} = \left(1 - K_n\right)^{1/2} \frac{\displaystyle\prod_{i=0}^{m_1}(y + y_i)\prod_{z=0}^{m_2}(y - y_z)}{q(y)}$$

$$= \frac{\displaystyle\prod_{j=0}^{m_2}(y - y_j)}{\displaystyle\prod_{j=0}^{m_2}(y + y_j)} \times \left(1 - K_n\right)^{1/2} \frac{\displaystyle\prod_{i=0}^{m_1}(y + y_i) \times \prod_{j=0}^{m_2}(y + y_j)}{q(y)} = \eta(s)\hat{\rho}(s)$$

where $y_i \geq 0$ and $y_j \geq 0$. Since

$$\rho(s) = \eta(s)\hat{\rho}(s) = 1 + \frac{\eta_1 + (\hat{b}_{n-1} - b_{n-1})\omega_c}{s} + \cdots$$

and since the coefficient constraints are given by

$$A_0 = \rho_0, \qquad \frac{A_1 - \rho_1}{F_2} \geq 0$$

we obtain

$$\left(1 - K_n\right)^{1/2} \geq \varepsilon \sinh\left\{ n \sinh^{-1}\left[\sinh\left(\frac{1}{n}\sinh^{-1}\frac{1}{\varepsilon}\right) - \frac{2\sin\dfrac{\pi}{2n}}{\omega_c}\left(\frac{1}{RC} - \sum_{i=1}^{u}\lambda_i\right) \right] \right\}$$

where

$$\sum_{i=1}^{u}\lambda_i \geq 0$$

From (4.57), we see that the term $\displaystyle\sum_{i=1}^{u}\lambda_i$ will increase the value on the right side of (4.57), thereby making K_n smaller. Thus, to maximize K_n, the bounded-real reflection coefficient is chosen to be

$$\rho(s) = \hat{\rho}(s) \qquad \text{or} \qquad \sum_{i=1}^{u}\lambda_i = 0$$

CHAPTER 5

THE ACTIVE LOAD

Problem 5.6 *Consider the same problem as in Example 5.1 except now that we wish to achieve the fifth-order Butterworth transducer power gain with passband tolerance of 3 dB. Realize the amplifier together with its schematic diagram.*

SOLUTION: Magnitude-scale the elements by a factor 10^{-2} and frequency-scale them by a factor 10^{-8}. Then

$$n = 5$$

$$G(\omega^2) = \frac{K}{1 + \omega^{2n}}$$

$$G(-s^2) = \frac{K}{1 + (-js)^{2n}} = \frac{K}{1 - s^{10}}$$

$$\frac{S_{12\beta}(s)S_{12\beta}(-s)}{S_{22\alpha}(s)S_{22\alpha}(-s)} = \frac{K}{1 - s^{10}}$$

Let

$$S_{12\beta}(s)S_{12\beta}(-s) = \frac{K}{K - s^{10}}$$

$$S_{22\alpha}(s)S_{22\alpha}(-s) = \frac{1 - s^{10}}{K - s^{10}}$$

Realize N_α as follows:

$$S_{22\alpha}(s)S_{22\alpha}(-s) = \frac{1 - s^{10}}{K - s^{10}} = \frac{1 - s^{10}}{\alpha^{10}\left[1 - \left(\dfrac{s}{\alpha}\right)^{10}\right]}$$

where $\alpha^{10} = K$. The minimum-phase factorization is found to be

$$\hat{S}_{22}(s) = \pm \frac{q(s)}{\alpha^5 q(s/\alpha)}$$

From the table on page 412 of the book, we obtain

$$\hat{S}_{22}(s) = \pm \frac{1 + 3.23607s + 5.23607s^2 + 5.23607s^3 + 3.23607s^4 + s^5}{\alpha^5 \left[1 + 3.23607\left(\frac{s}{\alpha}\right) + 5.23607\left(\frac{s}{\alpha}\right)^2 + 5.23607\left(\frac{s}{\alpha}\right)^3 + 3.23607\left(\frac{s}{\alpha}\right)^4 + \left(\frac{s}{\alpha}\right)^5 \right]}$$

$$= \pm \frac{1 + 3.23607s + 5.23607s^2 + 5.23607s^3 + 3.23607s^4 + s^5}{\alpha^5 + 3.23607\alpha^4 s + 5.23607\alpha^3 s^2 + 5.23607\alpha^2 s^3 + 3.23607\alpha s^4 + s^5}$$

$$z_1(s) = -1 + \frac{1}{\frac{2}{3}s - \frac{1}{3}} = -1 + \frac{3}{2s-1} = \frac{-2s+4}{2s-1}$$

$$z_3(s) = -z_1(-s) = \frac{2s+4}{2s+1}$$

$$r_3(s) = \frac{1}{2}\left(\frac{2s+4}{2s+1} + \frac{2s-4}{2s-1}\right) = \frac{1}{2} \times \frac{4s^2 + 8s - 2s - 4 + 4s^2 - 8s + 2s - 4}{(2s+1)(2s-1)}$$

$$= \frac{4s^2 - 4}{(2s+1)(2s-1)}$$

$$A(s) = \frac{s - \frac{1}{2}}{s + \frac{1}{2}} = \frac{2s-1}{2s+1}$$

$$\frac{r_3(s)}{z_3(s)} = \frac{4s^2 - 4}{(2s+1)(2s-1)} \times \frac{2s+1}{2s+4} = \frac{4(s+1)(s-1)}{(2s-1)(2s+4)}$$

Thus, $s_0 = 1$ is a Class I zero of transmission of order 1. The coefficient constraints become

$$\tilde{\rho}_0 = \tilde{A}_0 \;\; \rightarrow \;\; \hat{s}_{22\alpha}(1) = A(1)$$

$$A(1) = \frac{1}{3}$$

$$\hat{s}_{22\alpha}(1) = \frac{18.944}{\alpha^5 + 3.23607\alpha^4 + 5.23607\alpha^3 + 5.23607\alpha^2 + 3.23607\alpha + 1}$$

$$\alpha^5 + 3.23607\alpha^4 + 5.23607\alpha^3 + 5.23607\alpha^2 + 3.23607\alpha - 55.833 = 0$$

There must be at least one real root and it can be found by trial and error. For $\alpha = +1$, we have -39.889; and for $\alpha = +2$ we have $+97.249$. The desired α is found to be

$$\alpha = 1.4798$$

or

$$K = \alpha^{10} = 50.35$$

obtaining

$$F(s) = 2r_3(s)A(s) = \frac{8(s^2 - 1)}{(2s - 1)(2s + 1)} \times \frac{2s - 1}{2s + 1} = \frac{8(s^2 - 1)}{(2s + 1)^2}$$

$$Z_{22u}(s) = \frac{F(s)}{A(s) - \hat{S}_{22\alpha}(s)} - z_3(s)$$

$$= \frac{\dfrac{8(s^2 - 1)}{(2s + 1)^2}}{\dfrac{2s - 1}{2s + 1} - \dfrac{s^5 + 3.23607s^4 + 5.23607s^3 + 5.23607s^2 + 3.23607s + 1}{s^5 + 4.7887s^4 + 11.466s^3 + 16.967s^2 + 15.517s + 7.0955}} - \frac{2s + 4}{2s + 1}$$

$$= \frac{8s^6 + 44.099s^5 + 114.54s^4 + 180.70s^3 + 181.35s^2 + 106.07s + 24.379}{(2s + 1)\left(1.1053s^4 + 5.5405s^3 + 12.300s^2 + 14.659s + 8.0969\right)}$$

$$= \frac{4s^5 + 20.05s^4 + 47.245s^3 + 66.727s^2 + 57.311s + 24.379}{1.1053s^4 + 5.5405s^3 + 12.300s^2 + 14.659s + 8.0969}$$

$$Z_{22\alpha}(s) = 3.6189s + \cfrac{1}{0.40455s + \cfrac{1}{2.8194s + \cfrac{1}{0.18705s + \cfrac{1}{0.63985s + 3.0109}}}}$$

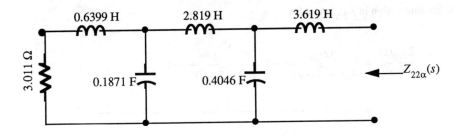

We next realize N_β, as follows:

$$S_{22\beta}(s)S_{22\beta}(-s) = 1 - S_{12\beta}(s)S_{12\beta}(-s) = 1 - \cfrac{K}{K - s^{10}}$$

$$= \cfrac{-s^{10}}{K - s^{10}} = \cfrac{\dfrac{-s^{10}}{K}}{1 - \dfrac{s^{10}}{K}} = \cfrac{-\left(\dfrac{s}{\alpha}\right)^{10}}{1 - \left(\dfrac{s}{\alpha}\right)^{10}}$$

where

$$K = \alpha^{10}$$

$$\hat{S}_{22\beta}(s) = \pm \cfrac{\left(\dfrac{s}{\alpha}\right)^{5}}{\left(\dfrac{s}{\alpha}\right)^{5} + 3.23607\left(\dfrac{s}{\alpha}\right)^{4} + 5.23607\left(\dfrac{s}{\alpha}\right)^{3} + 5.23607\left(\dfrac{s}{\alpha}\right)^{2} + 3.23607\left(\dfrac{s}{\alpha}\right) + 1}$$

$$= \pm \frac{s^5}{s^5 + 3.23607\alpha s^4 + 5.23607\alpha^2 s^3 + 5.23607\alpha^3 s^2 + 3.23607\alpha^4 s + \alpha^5}$$

$$= \pm \frac{s^5}{s^5 + 4.7887s^4 + 11.466s^3 + 16.967s^2 + 15.517s + 7.0955}$$

Choosing + sign in $\widehat{S}_{22\beta}(s)$ yields

$$\frac{Z_{22\beta}(s)}{R_2} = \frac{2A(s)}{A(s) - \widehat{S}_{22\beta}(s)} - 1 = \frac{2}{1 - \widehat{S}_{22\beta}(s)} - 1$$

$$= \frac{2}{1 - \dfrac{s^5}{s^5 + 4.7887s^4 + 11.466s^3 + 16.967s^2 + 15.517s + 7.0955}} - 1$$

$$= \frac{2s^5 + 4.7887s^4 + 11.466s^3 + 16.967s^2 + 15.517s + 7.0955}{4.7887s^4 + 11.466s^3 + 16.967s^2 + 15.517s + 7.0955}$$

$$Z_{22\beta}(s) = 0.41765s + \cfrac{1}{1.0934s + \cfrac{1}{1.3515s + \cfrac{1}{1.0932s + \cfrac{1}{0.41779s + 1}}}}$$

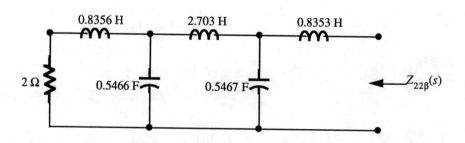

Magnitude-scaling the elements back by a factor of 10^2 and frequency-scaling them back by 10^8 yield the final design below:

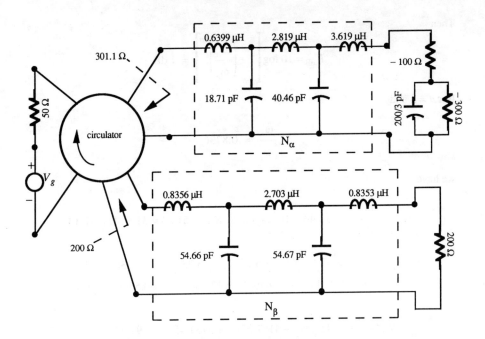

Problem 5.12 *Repeat the problem considered in Example 5.2 with the additional constraint that the variation inside the passband cannot exceed 1 dB.*

SOLUTION: Repeat Example 5.2 with

$$\text{passband ripple } R_{dB} \leq 1 \text{ dB}, \quad K_n \geq 30 \text{ dB}, \quad R_1 = 50 \ \Omega$$

$$R_2 = 200 \ \Omega, \quad -R = -100 \ \Omega, \quad C = 50 \text{ pF}, \quad \omega'_c = 10^8 \text{ rad/s}$$

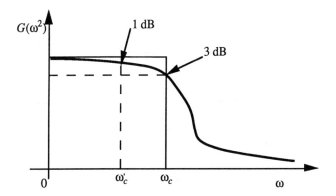

Then

$$R_{dB} = 10 \log \left[1 + \left(\frac{\omega'_c}{\omega_c} \right)^{2n} \right] = 1 \text{ dB}$$

For $n = 5$ and

$$\frac{\omega'_c}{\omega_c} = 0.8736$$

we have

$$\omega_c = \frac{\omega'_c}{0.8736} = 1.1447 \times 10^8 \text{ rad/s}, \qquad K_5 = 1514.4 = 31.8 \text{ dB} > 31 \text{ dB}$$

$$\hat{S}_{22\alpha}(s) = \frac{y^5 + 3.2361y^4 + 5.2361y^3 + 5.2361y^2 + 3.2361y + 1}{y^5 + 6.7305y^4 + 22.6494y^3 + 47.1064y^2 + 60.5506y + 38.9153}$$

$$Z_{22\alpha}(s) = \frac{349.44y^4 + 1741.33y^3 + 4187.03y^2 + 5731.45y + 3791.53}{3.92y^3 + 19.54y^2 + 42.09y + 39.92}$$

$$= 7.79 \times 10^7 s + \cfrac{1}{7.87 \times 10^{-11} s + \cfrac{1}{4.796 \times 10^{-7} s + \cfrac{1}{1.83 \times 10^{-11} s + \cfrac{1}{94.98}}}}$$

$$\left[\frac{Z_{22\beta}(s)}{200} \right]^{\pm 1} = \frac{2y^5 + 6.7305y^4 + 22.6494y^3 + 47.1064y^2 + 60.5506y + 38.9153}{6.7305y^4 + 22.6494y^3 + 47.1064y^2 + 60.5506y + 38.9153}$$

$$Z_{22\beta}(s) = 5.19 \times 10^{-7} s + \cfrac{1}{3.4 \times 10^{-11} s + \cfrac{1}{1.68 \times 10^{-6} s + \cfrac{1}{3.4 \times 10^{-11} s + \cfrac{1}{5.19 \times 10^{-7} s + \cfrac{1}{200}}}}}$$

$$R_1 = 50 \ \Omega, \qquad \hat{R}_2 = 200 \ \Omega, \qquad \hat{R}_3 = 94.98 \ \Omega$$

$$\mathbf{Y}_c(s) = 10^{-2} \begin{bmatrix} 0 & -1 & 1.45 \\ 1 & 0 & -0.726 \\ -1.45 & 0.726 & 0 \end{bmatrix}$$

The designed tunnel diode amplifier is given by

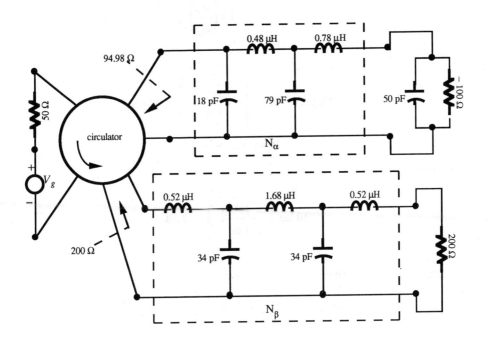

Problem 5.13 *Design a low-pass nonreciprocal amplifier for the following specifications:*

$$-R = -100\ \Omega, \quad C = 10\ pF$$
$$R_1 = -100\ \Omega, \quad R_2 = 200\ \Omega, \quad f_c = 100\ MHz$$

The amplifier is required to have a maximally-flat transducer power-gain characteristic with at least 30 dB within the passband and the passband tolerance is 2 dB.

SOLUTION:

$$G\left(\omega^2\right) = \frac{K_n}{1 + \left(\dfrac{\omega}{\omega_c}\right)^{2n}}$$

After substituting ω_c by ω'_c, we obtain

$$G\left(\omega'^2_c\right) = \frac{K_n}{1 + \left(\dfrac{\omega'_c}{\omega_c}\right)^{2n}}$$

Since

$$10 \log K_n - 10 \log\left[1 + \left(\frac{\omega'_c}{\omega_c}\right)^{2n}\right] = 40 \text{ dB}$$

$$10 \log K_n = 42 \text{ dB}$$

we have

$$10 \log\left[1 + \left(\frac{\omega'_c}{\omega_c}\right)^{2n}\right] = 2 \text{ dB}$$

Suppose that $n = 8$ and $\omega_c = 1$. Then we have

$$10 \log\left(1 + \omega'^{16}_c\right) = 2 \quad \rightarrow \quad 10^{2/10} = 1 + \omega'^{16}_c$$

$$\omega'^{16}_c = 10^{2/10} - 1 = 0.5848932$$

or

$$\omega'_c = 0.9670352$$

From

$$\omega'_c = 2\pi f'_c = 2\pi \times 10^8$$

$$\frac{\omega_c}{\omega'_c} = \frac{1}{\omega'_c} = \frac{n_{\omega_c}}{2\pi \times 10^8}$$

we obtain

$$n_{\omega_c} = \frac{2\pi \times 10^8}{\omega'_c} = \frac{2\pi \times 10^8}{0.9670352} = 2\pi \times 103.40885 \times 10^6 = 6.4973678 \times 10^8$$

Thus, when $\omega_c = 1$, $\omega'_c = 0.9670352$; and when $\omega'_c = 6.2831853 \times 10^8$,

$$\omega_c = 6.4973698 \times 10^8$$

Frequency- and magnitude-scaling the elements by the factors of

$$a = \frac{1}{6.4973698 \times 10^8} = 0.15391 \times 10^{-8}$$

$$b = 10^{-2}$$

respectively, yields $\omega_c = 1$, $R = 1$ and $C = 0.64973698$. Then

$$z_1(s) = \frac{1}{Cs - \dfrac{1}{R}} = \frac{1}{0.64973698s - 1}$$

$$z_3(s) = -z_1(-s) = \frac{1}{Cs + \dfrac{1}{R}} = \frac{1}{0.64973698s + 1}$$

$$r_3(s) = \frac{1}{2}\left[z_3(s) + z_3(-s)\right] = \frac{1}{2}\left[\frac{1}{Cs + \dfrac{1}{R}} + \frac{1}{-Cs + \dfrac{1}{R}}\right] = \frac{1}{\left(Cs + \dfrac{1}{R}\right)\left(\dfrac{1}{R} - Cs\right)}$$

$$= \frac{1}{(0.64973698s + 1)(1 - 0.64973698s)}$$

$$A(s) = \frac{s - \dfrac{1}{RC}}{s + \dfrac{1}{RC}} = \frac{RCs - 1}{RCs + 1} = \frac{Cs - 1}{Cs + 1} = 1 - \frac{2}{C} \times \frac{1}{s} + \frac{2}{C^2} \times \frac{1}{s^2} + \cdots$$

$$F(s) = 2r_3(s)A(s) = 2\frac{1}{(Cs + 1)(1 - Cs)} \times \frac{Cs - 1}{Cs + 1} = \frac{-2}{(Cs + 1)^2}$$

$$= 0 + \frac{0}{s} - \frac{2}{C^2} s^{-2} + \frac{4}{C^3} s^{-3} + \cdots$$

$$\frac{r_3(s)}{z_3(s)} = \frac{1}{(Cs+1)(1-Cs)} \times \frac{Cs+1}{1} = \frac{1}{1-Cs}$$

implying that $s_0 = \infty$ is a Class II zero of transmission of order $k = 1$. Therefore, we have the following relationships:

$$A_0 = 1, \quad A_1 = -\frac{2}{C}, \quad F_2 = -\frac{2}{C^2}$$

Suppose that

$$G(\omega^2) = \frac{K_n + \omega^{2n}}{1 + \omega^{2n}} = \frac{K_8 + \omega^{16}}{1 + \omega^{16}}$$

Then

$$\left| S_{22}^{(2)}(j\omega) \right|^2 = \frac{1 + \omega^{16}}{K_8 + \omega^{16}} = \frac{1 + \omega^{16}}{K_8 \left(1 + \dfrac{\omega^{16}}{\alpha^{16}} \right)} = \frac{1 + \omega^{16}}{\alpha^{16} \left(1 + \hat{\omega}^{16} \right)} = \frac{1 + s^{16}}{\alpha^{16} \left(1 + \hat{s}^{16} \right)} \Bigg|_{s-j\omega}$$

where

$$\alpha^{16} = K_8, \quad \hat{\omega} = \frac{\omega}{\alpha}, \quad \hat{s} = \frac{s}{\alpha}$$

from which we obtain the polynomial

$$q(y) = 1 + a_1 y + a_2 y^2 + a_3 y^3 + a_4 y^4 + a_5 y^5 + a_6 y^6 + a_7 y^7 + a_8 y^8$$

$$a_k = \prod_{u=1}^{k} \frac{\cos \dfrac{(u-1)\pi}{16}}{\sin \dfrac{u\pi}{16}}$$

yielding

$$a_0 = 1$$

$$a_1 = \frac{a_0 \cos 0}{\sin \pi/16} = 5.125830893$$

$$a_2 = \frac{a_1 \cos \pi/16}{\sin \pi/8} = 13.13707118$$

$$a_3 = \frac{a_2 \cos \pi/8}{\sin 3\pi/16} = 21.84615096$$

$$a_4 = \frac{a_3 \cos 3\pi/16}{\sin \pi/4} = 25.68835592$$

$$a_5 = \frac{a_4 \cos \pi/4}{\sin 5\pi/16} = 21.84615096$$

$$a_6 = \frac{a_5 \cos 5\pi/16}{\sin 3\pi/8} = 13.13707118$$

$$a_7 = \frac{a_6 \cos 3\pi/8}{\sin 7\pi/16} = 5.125830894$$

$$a_8 = \frac{a_7 \cos 7\pi/16}{\sin \pi/2} = 1$$

Therefore, we have

$$S_{22}^{(2)}(s) = \frac{s^8 + a_1 s^7 + a_2 s^6 + a_3 s^5 + a_4 s^4 + a_5 s^3 + a_6 s^2 + a_7 s + a_8}{s^8 + a_1 \alpha s^7 + a_2 \alpha^2 s^6 + a_3 \alpha^3 s^5 + a_4 \alpha^4 s^4 + a_5 \alpha^5 s^3 + a_6 \alpha^6 s^2 + a_7 \alpha^7 s + a_8 \alpha^8}$$

$$= 1 + \frac{a_1(1 - \alpha)}{s} + \dots$$

The coefficient constraints become

$$A_0 = 1, \quad s_0 = 1 \quad \rightarrow \quad A_0 = s_0$$

$$\frac{A_1 - s_1}{F_2} = \frac{-\frac{2}{C} - a_1(1 - \alpha)}{-\frac{2}{C^2}} \geq 0 \quad \rightarrow \quad \frac{2}{C} + a_1 \geq a_1 \alpha$$

showing that

$$\alpha \leq 1 + \frac{2}{a_1 C} \quad \rightarrow \quad \alpha \leq 1.600520915$$

or from

$$K_n \leq \left(1 + \frac{\sin \pi/2n}{\pi R C f_c}\right)^{2n}$$

we obtain

$$K_8 \leq \left[1 + \frac{2 \sin \pi/16}{\omega_c R C}\right]^{16} = \left[1 + \frac{2 \sin \pi/16}{0.64773678}\right]^{16}$$

yielding

$$K_8 \leq 1.600520915^{16} = 1854.3071 \quad \text{or} \quad 32.6818 \text{ dB}$$

Therefore, we choose $\alpha = 1.600520915$. Let

$$S_{22}^{(2)}(s) = \frac{x(s)}{y(s)}$$

The equalizer back-end impedance becomes

$$Z_{22}^{(2)}(s) = \frac{F(s)}{A(s) - S_{22}^{(2)}(s)} - z_3(s)$$

$$= \frac{\dfrac{-2}{(Cs+1)^2}}{\dfrac{Cs-1}{Cs+1} - \dfrac{x(s)}{y(s)}} - \frac{1}{Cs+1} = \frac{1}{1+Cs}\left[\frac{-y(1+Cs) + x(1+Cs)}{(y-x)Cs - (y+x)}\right]$$

$$= \frac{x-y}{(y-x)Cs - (y+x)} = \frac{y-x}{(x+y) - Cs(y-x)}$$

where

$$x + y = 2s^8 + 13.3298304s^7 + 46.78987551s^6 + 111.415412s^5 + 194.2589140s^4$$
$$+ 251.2927892s^3 + 233.9714455s^2 + 143.6351932s + 44.06166616$$

$$y - x = 3.078168668s^7 + 20.51573315s^6 + 67.72311014s^5 + 142.8822022s^4$$

$$+ 207.6004872s^3 + 207.6973032s^2 + 132.7835314s + 42.06166616$$

$$C = 0.649736977$$

The numerator of $Z_{22}^{(2)}(s)$ is $y - x$, whereas its denominator is

$$(x + y) - Cs(y - x)$$

or

$$0s^8 + 0s^7 + 2.787666650s^6 + 18.57956198s^5 + 59.373201s^4 + 116.3441713s^3$$

$$+ 147.6970752s^2 + 115.7061731s + 44.06166616$$

obtaining

$$Z_{22}^{(2)}(s) = \frac{3.08s^7 + 20.52s^6 + 67.72s^5 + 142.88s^4 + 207.60s^3 + 207.70s^2 + 132.78s + 42.06}{2.79s^6 + 18.58s^5 + 59.37s^4 + 116.34s^3 + 147.70s^2 + 115.71s + 44.06}$$

$$Z_{22}^{(2)}(s) = 1.104s + \cfrac{1}{1.289s + \cfrac{1}{1.083s + \cfrac{1}{0.998s + \cfrac{1}{0.683s + \cfrac{1}{0.464s + \cfrac{1}{0.143s + 0.955}}}}}}$$

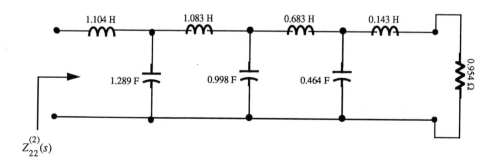

Another way to compute the element values from $Z_{22}^{(2)}(s)$ is as follows:

$$L_1 = L_c$$

$$C_{2m}L_{2m-1} = \frac{4 \sin \gamma_{4m-1} \sin \gamma_{4m+1}}{\omega_c^2 \left(1 - 2\delta \cos \gamma_{4m} + \delta^2\right)}, \quad m \le \frac{1}{2}(n-1)$$

$$C_{2m}L_{2m+1} = \frac{4 \sin \gamma_{4m+1} \sin \gamma_{4m+3}}{\omega_c^2 \left(1 - 2\delta \cos \gamma_{4m+2} + \delta^2\right)}, \quad m < \frac{1}{2}(n-1)$$

where

$$\gamma_m = \frac{m\pi}{2n}, \qquad \delta = K_n^{1/2n} = 1.600520915 \ \ (n = 8)$$

$$-\frac{L_c \omega_c}{R} = \frac{4 \sin \gamma_1 \sin \gamma_3}{C''_1 \left(1 - 2\delta \cos \gamma_2 + \delta^2\right)}$$

where $C''_1 = -RC\omega_c$, obtaining

$$L_c = -\frac{R}{\omega_c} \times \frac{4 \sin \gamma_1 \sin \gamma_3}{(-RC\omega_c)\left(1 - 2\delta \cos \gamma_2 + \delta^2\right)} = 1.104209739 = L_1$$

$$C_2 = \frac{4 \sin \gamma_3 \sin \gamma_5}{\omega_c^2 \left(1 - 2\delta \cos \gamma_4 + \delta^2\right) L_1} = 1.289008868$$

$$L_3 = \frac{4 \sin \gamma_5 \sin \gamma_7}{\omega_c^2 \left(1 - 2\delta \cos \gamma_6 + \delta^2\right) C_2} = 1.082991088$$

$$C_4 = \frac{4 \sin \gamma_7 \sin \gamma_9}{\omega_c^2 \left(1 - 2\delta \cos \gamma_8 + \delta^2\right) L_3} = 0.997538414$$

$$L_5 = \frac{4 \sin \gamma_9 \sin \gamma_{11}}{\omega_c^2 \left(1 - 2\delta \cos \gamma_{10} + \delta^2\right) C_4} = 0.683154177$$

$$C_6 = \frac{4 \sin \gamma_{11} \sin \gamma_{13}}{\omega_c^2 \left(1 - 2\delta \cos \gamma_{12} + \delta^2\right) L_5} = 0.464322608$$

$$L_7 = \frac{4 \sin \gamma_{13} \sin \gamma_{15}}{\omega_c^2 \left(1 - 2\delta \cos \gamma_{14} + \delta^2\right) L_6} = 0.14322899$$

$$\hat{R}_3 = R \frac{\delta^8 - 1}{\delta^8 + 1} = 0.954609070$$

$$a = \frac{1}{6.4973678 \times 10^8}, \qquad b = \frac{1}{100}$$

$$\omega_c = 1, \qquad L'_c = \frac{L_c a}{b} = 0.16795 \; \mu H$$

$$C'_2 = C_2 ab = 19.839 \; pF, \qquad L'_3 = \frac{L_3 a}{b} = 0.16668 \; \mu H$$

$$C'_4 = C_4 ab = 15.353 \; pF, \qquad L'_5 = \frac{L_5 a}{b} = 0.10515 \; \mu H$$

$$C'_6 = C_6 ab = 7.147 \; pF, \qquad L'_7 = \frac{L_7 a}{b} = 0.02204 \; \mu H$$

$$R = \hat{R}'_3 = \frac{\hat{R}_3}{b} = 95.461 \; \Omega$$

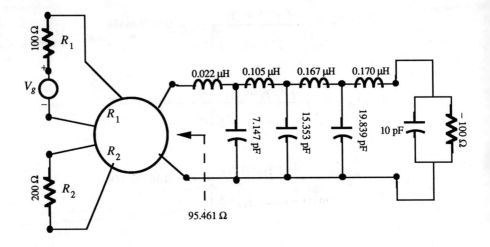

Problem 5.19 *Consider the same problem specified in Problem 5.13 except now that we wish to design an amplifier having an equiripple transducer power-gain characteristic, everything else being the same. Realize the amplifier together with its schematic diagram.*

SOLUTION: Given

$$-R = -100\ \Omega, \quad C = 10\ \text{pF}, \quad R_1 = 100\ \Omega, \quad R_2 = 200\ \Omega$$

and the Chebyshev response with

$$\omega_c = 2\pi \times 10^8\ \text{rad/s}, \quad K_n \geq 30\ \text{dB}, \quad \text{passband ripple } R_{dB} \leq 2\ \text{dB}$$

we obtain

$$\varepsilon = \sqrt{10^{0.2} - 1} = 0.76478$$

Choose $n = 4$. Then

$$K_4 = 33\ \text{dB} > 30\ \text{dB}, \quad \hat{\varepsilon} = \frac{\varepsilon}{\sqrt{K_4}} = 0.0171$$

$$\widehat{S}_{22\alpha}(s) = \frac{y^4 + 0.7162y^3 + 1.2565y^2 + 0.5168y + 0.2058}{y^4 + 3.8993y^3 + 8.6023y^2 + 11.2032y + 7.3097}$$

$$Z_{22\alpha}(s) = \frac{318.31y^3 + 734.59y^2 + 1068.64y + 710.396}{3.1444y^2 + 7.2564y + 7.5155}$$

$$= 1.61\times10^{-7}s + \cfrac{1}{1.63\times10^{-11}s + \cfrac{1}{6.52\times10^{-8}s + 94.5}}$$

$$S_{22\beta}(s) = \pm\frac{y^4 + y^2 + 0.125}{y^4 + 3.899y^3 + 8.602y^2 + 11.203y + 7.31}$$

$$\frac{Z_{22\beta}(s)}{R_2} = \pm\frac{2y^4 + 3.899y^3 + 9.602y^2 + 11.203y + 7.435}{3.899y^3 + 7.602y^2 + 11.203y + 7.185}$$

$$Z_{22\beta}(s) = 0.16\times10^{-6}s + \cfrac{1}{8.5\times10^{-12}s + \cfrac{1}{0.33\times10^{-6}s + \cfrac{1}{3.94\times10^{-12}s + \cfrac{1}{206.96}}}}$$

$$R_1 = 100\,\Omega, \quad \widehat{R}_2 = 206.96\,\Omega, \quad \widehat{R}_3 = 94.5\,\Omega$$

$$\mathbf{Y}_c(s) = 10^{-3}\begin{bmatrix} 0 & -6.95 & 10.3 \\ 6.95 & 0 & -7.15 \\ -10.3 & 7.15 & 0 \end{bmatrix}$$

The final design is shown below:

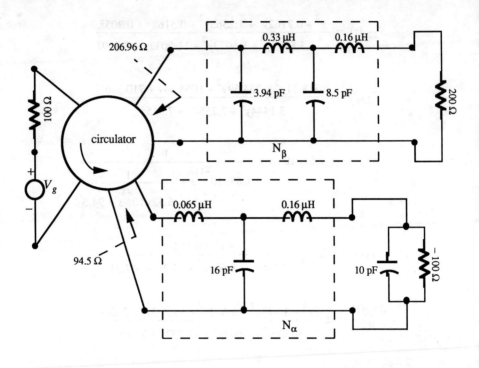

Problem 5.41 *Repeat the problem given in Example 5.4 for the sixth-order Butterworth transducer power-gain characteristic.*

SOLUTION: Given $n = 6$, the Butterworth response, and

$$-R = -143 \ \Omega, \quad C = 7 \ \text{pF}, \quad R_1 = 90 \ \Omega$$

$$R_2 = 500 \ \Omega, \quad f_c = 40 \ \text{MHz}, \quad K_6 \geq 37 \ \text{dB}$$

$$\hat{K}_6 = \left(\frac{\sin 15°}{\pi \times 0.401 \times 500 \times 7 \times 10^{-12} \times 4 \times 10} + 1 \right)^{12} = 50939.68 \quad \text{or} \quad 47.07 \ \text{dB}$$

$$\hat{S}_{22}(s) = \frac{y^6 + 3.86370y^5 + 7.46410y^4 + 9.14162y^3 + 7.46410y^2 + 3.86370y + 1}{2.467^6 \left(x^6 + 3.86370x^5 + 7.46410x^4 + 9.14162x^3 + 7.46410x^2 + 3.86370x + 1 \right)}$$

$$= \frac{y^6 + 3.8637y^5 + 7.4641y^4 + 9.1416y^3 + 7.4641y^2 + 3.8637y + 1}{y^6 + 9.5336y^5 + 45.4451y^4 + 137.3366y^3 + 276.6917y^2 + 353.408y + 225}$$

Using (4.93), the equalizer back-end impedance $Z_{22}(s)$ is found to be

$$Z_{22}(s) = 200.5 \frac{5.6699y^5 + 37.981y^4 + 128.195y^3 + 269.2276y^2 + 349.544y + 224.7}{7.69488y^4 + 51.52y^3 + 160.87y^2 + 278.02y + 226.7}$$

$$Z_{22}(s) = 5.88 \times 10^{-7}s + \cfrac{1}{1.58 \times 10^{-11}s + \cfrac{1}{5 \times 10^{-7}s + \cfrac{1}{7.56 \times 10^{-12}s + \cfrac{1}{1.42 \times 10^{-7}s + 198.73}}}}$$

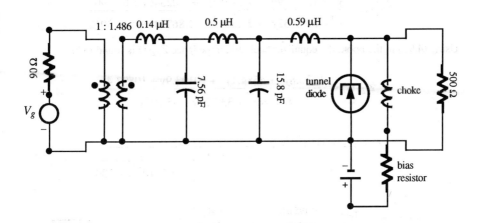

Problem 5.42 *Consider the same problem as in Example 5.4 except now that we wish to achieve an equiripple characteristic for its transducer power gain. What is the smallest n that will satisfy all the requirements? If n ≠ 5, realize the amplifier.*

SOLUTION: From specifications, we have

$$R_1 = 90 \ \Omega, \qquad R_2 = 500 \ \Omega, \qquad \omega_c = 2.513 \times 10^8 \ \text{rad/s}$$

$$-R = -143\ \Omega, \qquad C = 7\ \text{pF}, \qquad K_n \geq 37\ \text{dB}$$

For the Chebyshev response, $n = 4$ is the minimum order required. Thus, we choose

$$n = 4, \qquad \varepsilon = 0.34931\ \ (0.5\text{-dB ripple})$$

obtaining

$$\alpha = \frac{R}{R_2 - R} = 0.401, \qquad \alpha R_2 = 200.5\ \Omega$$

$$\widehat{K}_n = 23391.4\ \text{or}\ 43.69\ \text{dB}$$

$$K_n = \alpha\left(\widehat{K}_n - 1\right) = 9379.55\ \text{or}\ 39.72\ \text{dB} > (37 + 0.5)\ \text{dB}$$

$$\widehat{\varepsilon} = \widehat{K}_n^{-1/2}\ \varepsilon = 0.00228394$$

$$\widehat{S}_{22}(s) = \frac{y^4 + 1.19739y^3 + 1.71687y^2 + 1.02546y + 0.37905}{y^4 + 6.86732y^3 + 24.58y^2 + 51.86794y + 54.73012}$$

Using (4.93) in the book, the equalizer back-end impedance $Z_{22}(s)$ is found to be

$$Z_{22}(s) = \frac{1136.8y^3 + 4584.1y^2 + 10193.9y + 10897.4}{8.3648y^2 + 33.724y + 55.1092}$$

$$Z_{22}(s) = 5.41 \times 10^{-7}s + \cfrac{1}{1.23 \times 10^{-11}s + \cfrac{1}{1.95 \times 10^{-7}s + 200}}$$

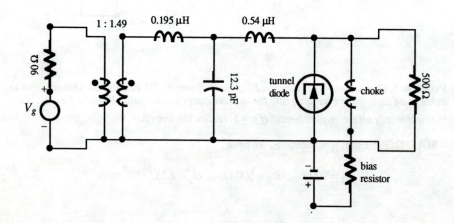

Problem 5.66 *Repeat Problems 5.13 and 5.19 for the tunnel diode which has the following specifications:*

$$-R = -50 \ \Omega, \quad C = 10 \ pF$$
$$R_d = 1 \ \Omega, \quad L_d = 10 \ nH$$

Also determine the resistive cutoff frequency f_r and the self-resonant frequency f_s of the tunnel diode.

SOLUTION: From the specifications, we have

$$R_1 = 100 \ \Omega, \quad R_2 = 200 \ \Omega, \quad f'_c = 10^8 \ Hz$$

$$K_n \geq 32 \ dB, \quad \text{passband ripple } R_{dB} \leq 2 \ dB$$

$$-R = -50 \ \Omega, \quad C = 10 \ pF = 10^{-11} F, \quad R_d = 1 \ \Omega, \quad L_d = 10 \ nH$$

obtaining

$$f_r = \frac{\sqrt{R/R_d - 1}}{2\pi RC} = 2228 \ MHz$$

$$f_s = \frac{\sqrt{\frac{R^2 C}{L_d} - 1}}{2\pi RC} = 389.8 \ MHz$$

Therefore, R_d and L_d can be omitted in the design.

(A) The Butterworth response

$$R_{dB} = 10 \log \left[1 + \left(\frac{\omega'_c}{\omega_c} \right)^{2n} \right] = 2 \ dB$$

Choose $n = 3$. Then

$$\omega_c = \frac{\omega'_c}{0.91449} = 6.8707 \times 10^8 \ rad/s$$

$$K_3 = \left(1 + \frac{2 \sin 30°}{5 \times 10^{-10} \omega_c} \right)^6 = 3578.2 = 35 \ dB > 32 \ dB$$

148

$$\widehat{S}_{22\alpha}(s) = \frac{y^3 + 2y^2 + 2y + 1}{y^3 + 7.8218y^2 + 30.5904y + 59.8181}$$

$$Z_{22\alpha}(s) = \frac{291.09y^2 + 1429.52y + 2940.91}{12.3843y + 60.8181}$$

$$= 3.42\times10^8 s + \cfrac{1}{6.13\times10^{-12}s + \cfrac{1}{48.36}}$$

$$\frac{Z_{22\beta}(s)}{200} = \frac{2y^3 + 7.8218y^2 + 30.5904y + 59.8181}{7.8218y^2 + 30.5904y + 59.8181}$$

$$Z_{22\beta}(s) = 7.4\times10^{-8}s + \cfrac{1}{3.7\times10^{-12}s + \cfrac{1}{7.4\times10^{-8}s + 200}}$$

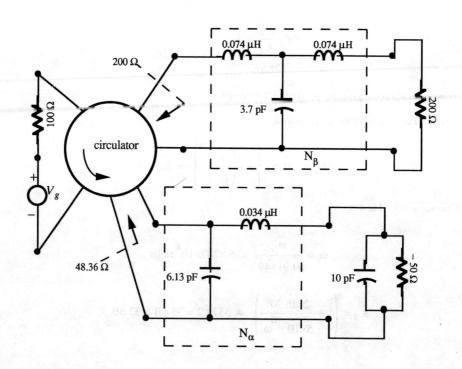

Also, we have

$$R_1 = 100 \ \Omega, \quad \hat{R}_2 = 200 \ \Omega, \quad \hat{R}_3 = 48.36 \ \Omega$$

$$\mathbf{Y}_c(s) = 10^{-3} \begin{bmatrix} 0 & -7.07 & 14.4 \\ 7.07 & 0 & -10.2 \\ -14.4 & 10.2 & 0 \end{bmatrix}$$

(B) The Chebyshev response

$$\omega_c = 2\pi \times 10^8 \ \text{rad/s}, \quad \varepsilon = \sqrt{10^{0.2} - 1} = 0.76478$$

Choose $n = 2$. Then we have

$$K_2 = 1605.9 \ \text{or} \ 32 \ \text{dB}$$

$$\hat{\varepsilon} = \frac{\varepsilon}{\sqrt{K_2}} = 0.01908$$

$$\hat{S}_{22\alpha}(s) = \frac{y^2 + 0.80382y + 0.82306}{y^2 + 7.1708y + 26.2102}$$

$$Z_{22\alpha}(s) = \frac{318.349y + 1269.357}{27.033} = 1.87 \times 10^8 s + 46.96$$

$$\hat{S}_{22\beta}(s) = \frac{y^2 + 0.5}{y^2 + 7.1708y + 26.2102}$$

$$\frac{Z_{22\beta}(s)}{R_2} = \frac{2y^2 + 7.1708y + 26.7102}{7.1708y + 25.7102} = 8.88 \times 10^{-8} s + \cfrac{1}{2.14 \times 10^{-12} s + \cfrac{1}{207.8}}$$

$$\mathbf{Y}_c(s) = 10^{-3} \begin{bmatrix} 0 & -6.937 & 14.59 \\ 6.937 & 0 & -10.12 \\ -14.59 & 10.12 & 0 \end{bmatrix}$$

where

$$R_1 = 100 \ \Omega, \quad \hat{R}_2 = 207.8 \ \Omega, \quad \hat{R}_3 = 46.96 \ \Omega$$

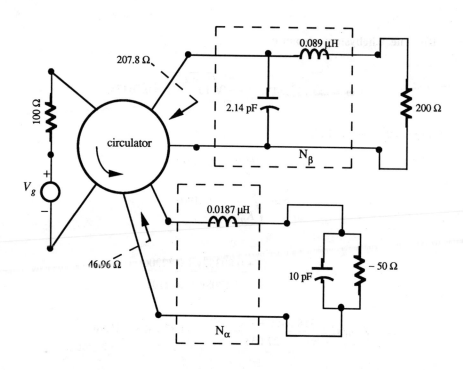

Problem 5.67 *Using the specifications given in Problem 5.13, design a transmission-power amplifier.*

SOLUTION: From the specifications, we have

$$-R = -100 \; \Omega, \quad C = 10 \; \text{pF}, \quad R_1 = 100 \; \Omega$$

$$R_2 = 200 \; \Omega, \quad f_c = 10^8 \; \text{Hz}, \quad K_n \geq 32 \; \text{dB}$$

obtaining

$$\alpha = \frac{R}{R_2 - R} = \frac{100}{200 - 100} = 1$$

Since

$$\lim_{n \to \infty} \hat{K}_n = \exp\left(\frac{1}{\alpha R_2 C f_c}\right) = \exp 5 = 148.4 \quad \text{or} \quad 21 \; \text{dB}$$

which is less than 32 dB, there is no solution.